*f*P

ALAS, POOR YORICK! I KNEW HIM,
HORATIO: A FELLOW OF INFINITE JEST, OF
MOST EXCELLENT FANCY: HE HAT BORNE
ME ON HIS BACK A THOUSAND TIMES;
AND NOW, HOW ABHORRED IN MY
IMAGINATIO

HAMLET
on the HOLODECK

THE FUTURE OF NARRATIVE
IN CYBERSPACE

Janet H. Murray

THE FREE PRESS

New York   London   Toronto   Sydney   Singapore

THE FREE PRESS
A Division of Simon & Schuster Inc.
1230 Avenue of the Americas
New York, NY 10020

Designed by Carla Bolte

Manufactured in the United States of America

10   9   8   7   6   5   4   3   2   1

Library of Congress Cataloging-in-Publication Data
Murray, Janet Horowitz, 1946–
    Hamlet on the holodeck : the future of narrative in cyberspace /
Janet H. Murray.
      p. cm.
    Includes index.
    1. Interactive multimedia. 2. Virtual reality. 3. Literature—
History and criticism. 4. Narration (Rhetoric) I. Title.
QA76.76.I59M87   1997
809'.00285'67—dc21   97-9187
                                                CIP

ISBN 0-684-82723-9

For my son, William

Contents

PART IV. NEW BEAUTY, NEW TRUTH

Acknowledgments

I have many people to thank for their generous aid in writing this book.

This is very much the book of someone who has spent the past twenty-five years at MIT, and I must begin by thanking those that I have worked with and learned from.

First of all, I am grateful to my students. Some of those in the interactive fiction writing course I have taught since 1992 are mentioned by name in the text of this book, but many more of them contributed imaginative and inventive projects that helped push my thinking about the new medium. I am particularly grateful to the graduate students whose theses I helped advise, including Ayshe Farman-Farmaian, Mark Halliday, Kevin Brooks, and Bradley Rhodes. I also learned a lot from working with Freedom Baird, Lee Morgenroth, David Kung, Michael Murtaugh, Richard Lachman, and Dave Tames. Most of these graduate students were members of Glorianna Davenport's Interactive Cinema Group at the Media Lab, and I am grateful to her for including me in their work. I am grateful as well to Jeffrey Morrow and Matthew Gray for programing versions of the Character Maker/Conversation program discussed in chapter 8.

I feel very fortunate to have been able to spend time with members of the Assassin's Guild, the virtuoso role-playing group at MIT, who graciously let me observe some of their games. In particular, I want to thank Seth McGinnis for introducing me to the Guild, and Andrea Humez for letting me see how an expert gamemaster works.

My work on digital media has been supported by three MIT deans, Harold Hanham, the late Ann Friedlaender, and Philip Khoury, and it was made possible by the help of my colleagues in the School of Engineering and the Media Lab, who were always ready to think about making something new and useful. My thinking about the aesthetics of the medium has been enriched by the process of designing software for humanities education, and I am grateful to all of those I have worked with in that effort. As one MIT engineer is fond of saying, the early Christians get the best lions, and we were very early into the arena and still have the wounds to show for it. I want to thank all the participants in and advisors to the Athena Language Learning Project, Berliner sehen, the Shakespeare Electronic Archive, and the Virtual Screening Room for the privilege of working with them on interactive design. I am particularly grateful to Douglas Morgenstern, who first suggested to me that we make interactive video narratives from the simulations he ran in his foreign language learning class, and who has been an unceasing source of creative ideas and friendship through almost fifteen years of collaborative work.

One of the great privileges of working in humanities computing in the past two decades has been my friendship with Larry Friedlander of Stanford University. The chapter "Transformation" owes much to my conversations with him, but more than that, his generous imagination has been a continuing source of inspiration to me.

I presented many of the core ideas in this book at conferences with participants who ranged from English teachers to computer scientists, and I have benefited from the thoughtful and energetic responses I received on all of those occasions. I am particularly grateful to have participated in the symposia Believable Characters (1994) and Interactive Story Systems (1995), both organized by Joseph Bates for the American Association for Artificial Intelligence, and in the Lifelike Computer Characters Conference of 1995. I also gained much from participation in the Future of Media Studies conference held at MIT in October 1995 and the Computers and Humanities Workshop held at MIT in May 1994. In addition I am grateful for having the chance to present my ideas at Mitsubishi Electric Research Laboratory, the Modern Language Association, the National Council of Teachers of English, the Association for Computers in the Humanities, the Paris conference Littérature Généré par Ordinateur, and a 1995 NEH Summer Institute on hypertext directed by Jay Bolter and Michael Joyce.

Most importantly, I owe a very large debt to those who graciously read large parts of the manuscript, often at very short notice: especially Norman Holland, Henry Jenkins, Sherry Turkle, and Peter Petre; and also Amy Bruckman, Bernice Buresh, Ann Banks, Glorianna Davenport, Tom Englehardt, Lenny Foner, Bradley Rhodes, Scott Reilly, and Harriet Rosenstein. I have been saved from many errors and confusions by their help. Whatever faults remain are entirely my own.

I also wish to thank those who took the time to answer crucial questions or who helped me to sort out my ideas in key conversations, including Hal Abelson, Hal Barwood, Joseph Bates, Robert Berwick, Jeffrey Bigler, Jay Bolter, Gregory Crane, Peter Donaldson, Steve Ehrmann, Clark Elliott, Sue Felshin, Richard Finneran, Ken Haas, Nick Hildebidle, David Jones, Noah Jorgensen, Michael Joyce, George Landow, Brenda Laurel, Steve Lebrande, Steven Lerman, Michael Malone, Stuart Malone, Kenneth Mayer, Ruth Perry, Barbara Sirota, Vivian Sobchack, David Thorburn, Lily Tomlin, Jane Wagner, Joseph Weizenbaum, Catherine White, Patrick Winston, and Gerald Wyckoff.

A book about a new medium needs the guidance of an editor who understands both books and bits. I am grateful for the intelligence and dedication of my editor Bruce Nichols at the Free Press and for the meticulous production work of Loretta Denner and Toby Troffkin. I particularly appreciate their patience in allowing me much disruptive revision while still insisting on getting the book out into the world. I also wish to thank my agent Charlotte Sheedy for her perceptive suggestions and energetic support of the project from the very beginning.

Finally, it is a joy to me to thank my family, whose love and encouragement made this book possible. In addition to tolerating my writer's self-absorption, they all volunteered as research assistants for the project. My mother, Lillian Horowitz, combed the newspapers and television coverage, providing me with many valuable bulletins, while demonstrating that the promise of the future can be just as thrilling in one's eighties as it is in one's teens. My husband Tom brought me countless cartons of library books, photocopied mountains of manuscripts, and was always willing to take one more walk while listening to my obsessive reformulations. His love and wit sustained me in this effort, as in everything I do. My daughter Elizabeth frequently took time from her own creative work to cheer me on and to counsel me from her perspective as an actress. Her artistry and courage are a continuing joy and inspiration to me.

But most of all, I want to thank my talented son, William, now sixteen years old, who has generously shared with me his delight in multiform narratives of every kind, who has educated me about the art of the comic book and the delights of the videogame, who has served as my trusty Internet sleuth, and whose bountiful imagination and keen literary intelligence were my constant companion through all the labyrinthian tangles of this investigation. I offer this book in dedication to him, with love and admiration.

Introduction: A Book Lover Longs for Cyberdrama

ALAS, POOR YORICK! I KNEW HIM, HORATIO: A FELLOW OF INFINITE JEST, OF MOST EXCELLENT FANCY: HE HATH BORNE ME ON HIS BACK A THOUSAND TIMES; AND NOW, HOW ABHORRED IN MY IMAGINATION IT

All media as extensions of ourselves serve to provide new transforming vision and awareness.

—Marshall McLuhan

Our various improvements not only mark a diminution of the function improved upon . . . but they also work to dissolve some of the fundamental authority of the human itself. We are experiencing the gradual but steady erosion . . . of the species itself.

—Sven Birkerts

The birth of a new medium of communication is both exhilarating and frightening. Any industrial technology that dramatically extends our capabilities also makes us uneasy by challenging our concept of humanity itself. (Are people meant to move across the ocean like the fish? Are people's words supposed to be transmitted by dead paper or cold wires?) The boat, car, and airplane are seemingly magi-

1

cal extensions of our arms and legs; the telephone extends our voices; and the book extends our memory. The computer of the 1990s, with its ability to transport us to virtual places, to connect us with people at the other end of the earth, and to retrieve vast quantities of information, combines aspects of all of these. And as if that were not amazing enough, it also runs our warplanes and plays a masterly game of chess. It is not surprising, then, that half of the people I know seem to look upon the computer as an omnipotent, playful genie while the other half see it as Frankenstein's monster. To me—a teacher of humanities for the past twenty-five years in the world-class electronic toy shop of MIT, a Victorian scholar, and educational software designer—the computer looks more each day like the movie camera of the 1890s: a truly revolutionary invention humankind is just on the verge of putting to use as a spellbinding storyteller.

It is somewhat surprising to me to find myself on the optimistic side of this pervasive new cultural device. When I first trained as a systems programmer, as an IBM employee in the late 1960s, I was only biding my time and saving up money for graduate school in English literature. I found the clean logic of computer programming satisfying, and I enjoyed deciphering the mysterious 0's and 1's of a "core dump" to reveal what the machine was up to when a program crashed (as they so often did). But there seemed no deeper purpose in this work than there had been in the intriguing geometry proofs I had enjoyed in high school and then promptly forgotten. For me at the age of twenty, the only activity worthy of serious human effort was reading novels.

Only once during my time at IBM did I catch a glimpse of a more inspired use of the computer. Although we did not use the terms at that time, the corporate world was clearly divided into "suits" and "hackers." The suits were running the company (better than they would in later years), but the hackers were running the secret playground within the company, the world of the machines. Computer systems in those days were mammoth arrays of cumbersome appliances kept isolated in ice-cold rooms. The tape drives alone (the equivalent of today's floppy disks) were the size of refrigerators. The noisiest component was the card reader, which jangled and thumped like a subway train full of bowling balls as it processed stacks of the punch cards that were the main form of human-to-computer communication in that era. Dealing with this machine was an unpleasant daily necessity. But one day the icy, clamorous cardprinter room was turned into a whimsical cabaret: a clever young hacker had created a set of punch cards that worked like a player piano roll and caused the card reader to chug out a recognizable version of the Marine Corps Hymn: bam-bam-THUMP bam-THUMP bam-THUMP THUMP-THUMP. All day long, programmers sneaked away from their work to hear this thunderously awful but mesmerizing concert. The data it was processing was of course meaningless, but the song was a work of true virtuosity.

When programming was fun, it was a lot like that performance. Creating a successful machine code program made me feel as if I had communicated with some recalcitrant, stupid beast deep inside the refrigerator cabinet and taught it a new little tune. But my real work was waiting for me somewhere else, in the form of a long, thoughtful walk down an endless shelf of books. When I was offered a fellowship for graduate school at Harvard, I did not hesitate to accept it. My IBM manager wanted me to take just a temporary leave of absence. He gave me an article about how computers were being used to study English literature (someone was putting all of *War and Peace*—to me the pinnacle of human wisdom—into electronic form in order to count the number of words in each of Tolstoy's sentences). The article ended by referring to literature as "man's greatest output." I told my manager to write me up as a permanent resignation.

I began reading my own way down that long shelf of books. I agreed with D. H. Lawrence that the novel was the one "bright book of life,"[1] the measure of all things, although I much preferred the work of Jane Austen and the Victorians. My favorite critic was Northrop Frye, who combined detailed analyses of the structure of stories with a profound appreciation of their mythic power. Reading

Frye it was possible to believe that the formal beauty of literary art is an expression of its deeper truth. Yet the more I read, the clearer it became that stories did not tell the whole truth about the world. As I researched the lives of women in the Victorian era, I (like others of my generation) was struck by the fact that much of what I was learning had been left out of the great novels of the era. Although my faith in the deeper powers of literature was unshaken, I learned from the feminist movement that some truths about the world are beyond the reach of a particular art form at a particular moment in time. Before the novel could tell the stories of women who did not wind up either happily married or dead, it would have to change in form as well as in content.

For the stories I wanted to hear, I looked in other formats, in feminist magazines and maverick novels.[2] I compiled an anthology documenting the experiences of Victorian women—prostitutes, medical students, circles of women friends—who had not found a place in classic fiction.[3] But the anthology format was as limiting in its way as the marriage plot. Frustrated by the constraint of producing a single book with a single pattern of organization, I filled my collection with multiple cross-references, encouraging the reader to jump from one topic to another. I simply wanted the reader to understand Mary Taylor's exhilaration in opening a dry goods store in New Zealand in the context of her friendship with Charlotte Brontë as well as in relation to the range of Victorian opinion on women's work. I did not think of this cross-referencing as hypertext because I had not yet heard the term.

Though I had been teaching at MIT since 1971, I was not drawn to computers again until the early 1980s. While I had been exploring social history and raising my two children, literature and academic feminism itself seemed somehow to have fallen into the hands of the suits. The new theoreticians no longer saw the novel as the "bright book of life" but as an infinite regression of words about words about words. Joining in this conversation involved learning a discourse as arcane as machine code, and even farther from experience. Truth and beauty were nowhere in sight. But at the same time that literary theorists were denouncing meaning as something to be deconstructed into absurdity, theorists of learning methods were embracing meaning as the key to successful pedagogy. One conference paper after another celebrated the fact that students wrote better papers and learned to speak foreign languages with greater fluency when they actually had something they wanted to communicate to one another.[4] The new research in cognition and sociolinguistics seemed to define what those processes of communication entailed. Thinking about teaching was much more satisfying to my earnest Victorian temperament than thinking about literary criticism. And the more I thought about it, the more I wondered if these practical and process-oriented methodologies could be transported into the world of the computer.

I was at that time the humanities faculty member in the Experimental Study Group (ESG), in which conventional courses were taught in an individualized manner. ESG attracted some of the most creative and self-directed students at MIT, many of whom were also ingenious computer hackers. They wrote their papers on-line, explored imaginary dungeons filled with trolls, exchanged wisecracks with computer-based imaginary characters, and engaged in a perpetual telnet tour of the globe by playfully breaking into other people's computers. They believed the particular programming language they were learning was both the brain's own secret code and a magical method for creating anything on earth out of ordinary English words.[5] They saw themselves as wizards and alchemists, and the computer as a land of enchantment. MIT was paradise for these hackers, who were largely engaged in navigating through an elaborate fictional universe. With such students as my guides, I got myself a network account and renewed my acquaintance with the digital world.

I had left computing in the age of punch cards and came back to it in the age of video display terminals and microcomputers. Nevertheless, educational computing had not advanced very far beyond the days of quantifying Tolstoy's "output." The computer was mostly seen as a drudge, a workhorse for word frequency analysis and for drill and practice teaching. However, to my students and my MIT colleagues, it was clearly something considerably more nimble. Seymour Papert had developed the LOGO programming language that allowed children to learn mathematical concepts by choreographing the actions of magic sprites that raced across the screen. A follower of Piaget, Papert believed that computers are tools for thinking and should be used to create "microworlds" where inquisitive students can learn through a process of exploration and discovery.[6] Nicholas Negroponte's group had created a suite of dazzling demonstration projects (the seed work for the Media Lab) that included a "movie map" of Aspen, Colorado, and a "movie manual" for car repair.[7] The combination of text, video, and navigable space suggested that a computer-based microworld need not be mathematical but could be shaped as a dynamic fictional universe with characters and events.

My interest in creating narrative microworlds coincided with the interests of foreign language teachers in creating immersive learning environments. Together we designed multimedia applications for learning Spanish and French, which motivated students by giving them a role in an unfolding story and allowing them to move through authentically photographed environments as if they were on a visit to Bogota or Paris.[8] These projects and others that I have worked on in the past fifteen years—including a Shakespeare archive and a film art digital textbook—as well as many kindred efforts pursued by others elsewhere, have confirmed my view of the computer as offering a thrilling extension of human powers. I say this despite the often agonizing uncertainties of software development and the continual frustration over the gap between what designers want the hardware and software to do and what they actually support.[9] For my experience in humanities computing has convinced me that some kinds of knowledge can be better represented in digital formats than they have been in print. The knowledge of a foreign language, for instance, can be better conveyed with examples from multiple speakers in authentic environments than with lists of words on a page. The dramatic power of Hamlet's soliloquies is better illustrated by multiple performance examples in juxtaposition with the text than by the printed version alone. Discussions of film art make more sense when they are grounded by excerpted scenes from the movies being discussed. Computers can present the text, images, and moving pictures valued by humanistic disciplines with a new precision of reference; they can show us all the different ways a French person says "hello" in a single day or all the passages Zeffirelli chose to leave out of his production of *Romeo and Juliet.* By giving us greater control over different kinds of information, they invite us to tackle more complex tasks and to ask new kinds of questions. Although the computer is often accused of fragmenting information and overwhelming us, I believe this view is a function of its current undomesticated state. The more we cultivate it as a tool for serious inquiry, the more it will offer itself as both an analytical and a synthetic medium.

My experiences in educational computing have also offered me evidence of how frightening the new technologies can be. Several years ago I was invited to talk with the committee that was then overseeing the production of a variorum Shakespeare, a set of editions of individual plays with extensive annotation covering all known textual variants as well as notes on the significant critical commentary on the plays.[10] The variorum format dates back to the nineteenth century and was still an endearingly Victorian endeavor. The pace of production was glacial, with many of the editors collecting their notes in stacks of index cards and filling hundreds of shoe boxes with twenty years' worth of investigation before publishing. The night before my appearance I was invited for a drink in a high-rise New York hotel room by two of the most computer-friendly committee members. I had already received an irate note from another member of the committee, and my hosts, an English woman and a Southern man, were anxious to prepare me for the kind of opposition that others might offer. My scrupulously polite colleagues displayed a courtly commitment to moving the variorum into the digital age while avoiding offending anyone. With the naïveté of someone who had spent much of the past twenty years in the company of engineers, I told them that my remarks would be limited to the obvious practicalities of their work. Clearly, the pages of a book were a poor match for the task at hand. Often the text of the play took up only a single line at the top, with the rest of the page covered with footnotes in several numbering schemes, many of which were condensed to cryptic abbreviations that conveyed no information to the uninitiated. Thus, commentary for a line of text often appeared a dozen pages away from the line it referred to. The effort of compiling a variorum edition was clearly heroic, but the arbitrary limitation of the printed page was a disservice to the depth of information and expertise involved. At this point in my preview of the next day's presentation, my genteel hostess started to shake in her chair. "I love the book!" she cried. "If you are coming to talk against the book tomorrow, I will throw you out the window." And though she was considerably smaller than I, she looked quite prepared to do so.

Why should the prospect of a scholarly CD-ROM bring a mild-mannered Shakespearean editor to such paroxysms of violence? To my mind it was because she could not separate the activities of research from the particular form they had historically assumed. Her love of books (which I share) momentarily blinded her to the true object of reverence: the creation of a superb reference work. Her reaction was a sign that the new technologies are extending our powers faster than we can assimilate the change. Even when we are already engaged in enterprises that cry out for the help of a computer, many of us still see the machine as a threat rather than an ally. We cling to books as if we believed that coherent human thought is only possible on bound, numbered pages.

I am not among those who are eager for the death of the book, as I hope the present volume demonstrates. Nor do I fear it as an imminent event. The computer is not the enemy of the book. It is the child of print culture, a result of the five centuries of organized, collective inquiry and invention that the printing press made possible.

My work as a software developer has made me painfully aware of the primitive nature of the current digital medium and of the difficulty of predicting what it can or cannot do in any given time scheme. Nevertheless, I find myself longing for a computer-based literary form even more passionately than I have longed for computer-based educational environments, in part because my heart belongs to the hackers. I am hooked on the charm of making the dumb machines sing.

Since 1992 I have been teaching a course on how to write electronic fiction. My students include freshmen, writing majors, and Media Lab graduate students. Some of them are virtuoso programmers. Some of them do not program at all. All of them are drawn to the medium because they want to write stories that cannot be told in other ways. These stories cover every range and style, from oral histories to adventure tales, from the exploits of comic book heroes to domestic dramas. The only constant in the course is that every year what is written is even more inventive than what was written the year before. Every year my students arrive in class feeling more at home with electronic environments and are more prepared to elicit something with the tone of a human voice out of the silent circuitry of the machine.

As I watch the yearly growth in ingenuity among my students, I find myself anticipating a new kind of storyteller, one who is half hacker, half bard. The spirit of the hacker is one of the great creative wellsprings of our time, causing the inanimate circuits to sing with ever more individualized and quirky voices; the spirit of the bard is eternal and irreplaceable, telling us what we are doing here and what we mean to one another. I am drawn to imagining a cyberdrama of the future by the same fascination that draws me to the Victorian novel. I see glimmers of a medium that is capacious and broadly expressive, a medium capable of capturing both the hairbreadth movements of individual human consciousness and the colossal crosscurrents of global society. Just as the computer promises to reshape knowledge in ways that sometimes complement and sometimes supersede the work of the book and the lecture hall, so too does it promise to reshape the spectrum of narrative expression, not by replacing the novel or the movie but by continuing their timeless bardic work within another framework.

This book is an effort to imagine what kinds of pleasures such a cyberliterature will bring us and what sorts of stories it might tell. I believe that we are living through a historic transition, as important to literary history as it is to the history of information processing. My sixteen-year-old son will no doubt look back upon the moment at which we (finally!) got our home computer hooked up to the World Wide Web with the same delight with which my father recalled plucking voices out of the air with his home-made crystal radio set. My paternal grandmother, who started life in a Russian shtetl, jumped in terror when she heard that disembodied speech, thinking it must be a dybbuk or ghost. Yet only a few decades later, I sat in my crib, as my mother fondly reports, calmly enraptured by the voice of Arthur Godfrey. Today, my husband collects tapes of old Bob and Ray programs, which we listen to on long car trips, savoring the intimacy of what now seems like a touchingly low-tech format. Those of us who have spent our lives in love with books may always approach the computer with something of my grandmother's terror before the crystal radio, but our children are already at home with the joystick, mouse, and keyboard. They take the powerful sensory presence and participatory formats of digital media for granted. They are impatient to see what is next. This book is an attempt to imagine a future digital medium, shaped by the hacker's spirit and the enduring power of the imagination and worthy of the rapture our children are bringing to it.

PART I

A New Medium for Storytelling

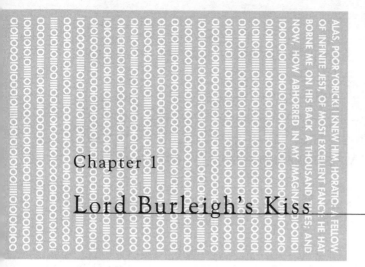

Chapter 1

Lord Burleigh's Kiss

In a far distant corner of the galaxy sometime in the twenty-fourth century, the brisk and competent Kathryn Janeway, captain of the starship *Voyager*, is taking a break from her duties with her favorite "holonovel." Exchanging her spandex-sleek Starfleet uniform for a hugely crinolined Victorian dress, Janeway enters one of the ship's "holosuites," which is running a three-dimensional simulation of a richly furnished English drawing room, complete with cozy armchairs and a roaring fire. Brooding by the fire is the handsome romantic hero, who greets her as she enters as his governess, Lucy Davenport. He gives her a meaningful look, and she returns it earnestly.

"Lord Burleigh, is something wrong?"

"Yes, terribly wrong."

He suddenly steps toward her, takes her in his arms, and kisses her passionately. "I have fallen in love with you, Lucy." They stare deeply into one another's eyes.

But it is teatime, and they are interrupted by the arrival of the sinister housekeeper and Lord Burleigh's two anxious and secretive young children. His little daughter, Beatrice, drops her teacup with alarm when questioned about the mysterious piano music that Lucy has been hearing.

Beatrice's precocious brother, Henry, is quick to silence her.

As soon as they are again alone, Lucy confronts Lord Burleigh: "What's happening in this house? How can you not know that Beatrice plays the piano? Why shouldn't I go to the fourth floor? What's up there?"

"Those are questions you must not ask," he declares imperiously.

"But I am asking them," comes her fervent reply. "I'm worried about the children. Beatrice fantasizes that her mother is still alive."

"Don't pursue this, I beg you," he says, looking deep into her eyes.

The confrontation is escalating dramatically and Lucy is breathless with excitement when suddenly another voice is heard:

"Bridge to Captain."

"Freeze program," says Lucy/Janeway, reluctantly backing away from the now frozen image of Lord Burleigh. "Janeway here."

"We've been hailed by a representative of the Bothan government. They'd like to talk to you."

"I'll be right there."

As she turns to leave, Janeway pauses before the stilled hologram of her would-be lover. "Sorry, my lord. Duty calls," she says, grinning, before striding back to resume command of the ship.[1]

Captain Janeway's Victorian excursion takes place on *Star Trek: Voyager,* the latest of four *Star Trek* television series in which gloriously equipped starships and space stations serve the ideals of the peace-seeking interplanetary United Federation of Planets.[2] There are many technical wonders in the *Star Trek* vision of the future, including lightspeed travel; photon weapons; medical "tricorders," which diagnose and heal with the wave of a wand; the well-known "transporter" room, in which technicians "beam" the crew up and down from dangerous planets by decomposing and reassembling their molecular patterns; and the conveniently wall-mounted "replicators," which can materialize hot and cold snacks on demand. The holodeck is an appropriate entertainment medium for the fortunate citizens of such a world: a utopian technology applied to the age-old art of story-telling.[3]

First introduced on *Star Trek: The Next Generation* in 1987, the holodeck consists of an empty black cube covered in white gridlines upon which a computer can project elaborate simulations by combining holography with magnetic "force fields" and energy-to-matter conversions. The result is an illusory world that can be stopped, started, or turned off at will but that looks and behaves like the actual world and includes parlor fires, drinkable tea, and characters, like Lord Burleigh and his household, who can be touched, conversed with, and even kissed. The *Star Trek* holodeck is a universal fantasy machine, open to individual programming: a vision of the computer as a kind of storytelling genie in the lamp. In the three series in which the holodeck has been featured, crew members have entered richly detailed worlds, including the tribal manor house of the Old English *Beowulf* saga, a gaslit London street, and a San Francisco speakeasy, in order to participate in stories that change around them in response to their actions.[4]

Lucy Davenport (as we can call Janeway's unnamed adventure) is in many ways typical of the holonovel form. It is a period piece and a work of genre fiction in which the elaborate set design and recognizable story conventions (an arrival in the rain, ghostly noises at the window, a forbidden attic) are playfully savored, as if put there by a very thorough and well-read programmer. Holonovels provide customized entertainment for a variety of tastes. They reveal unexpected sides of familiar characters. Just as Jean-Luc Picard, the highly cultured captain of *Star Trek: The Next Generation,* enjoys film noir, his android crewman, Commander Data, identifies with Sherlock Holmes, and the sensitive Dr. Julian Bashir of *Star Trek: Deep Space Nine* prefers James Bond spy adventures, so the conscientious Captain Janeway turns to gothic fiction in her well-earned leisure hours.

But Janeway's holonovel marks a milestone in this virtual literature of the twenty-fourth century as the first holodeck story to look more like a nineteenth-century novel than an arcade shoot-'em-up. Unlike virtually all the holodeck stories run by male crew members, *Lucy Davenport* is not focused on a violent central conflict that is resolved within a single Star Trek episode. Instead, Janeway is involved in a more leisurely and open-ended exploration of the Burleigh household, a continuing avocation that she takes up regularly on her days off and that is presented over several episodes.[5] From Janeway's references to events that are not dramatized, it seems that she is spending long periods of time in the household, participating in a daily routine, giving lessons to the children, having tea at regular hours, and getting to know each individual character. Like *Jane Eyre*, Charlotte Brontë's 1847 novel, which established the governess gothic genre, *Lucy Davenport* takes place in a mysteriously haunted household and emphasizes the perils of the governess's intense social relationships rather than the physical terrors of the situation. When Janeway is shown relishing a verbal contest with the sinister housekeeper, promising the reluctant Henry that she will be a challenging math teacher, or trying to assuage the grief of the clearly anguished young Beatrice, we can understand what engages the resourceful starship captain in this particular virtual world. As her name implies, Janeway has much in common with her fictional predecessor Jane Eyre, including a strong resistance to being bullied, a willingness to stand on principle, and the courage to face fear and isolation head-on. The Lucy Davenport story therefore suits her well, making the holodeck form itself seem worthy of adult attention.

Janeway's attraction to the illusory Lord Burleigh is taken seriously as an exercise posing psychological and moral questions for her. After she is surprised by his teatime embrace, Janeway is tormented by visions of the holonovel characters walking around the ship. She thinks she is hallucinating until it is discovered that a telepathic enemy alien is fabricating these visions as a way of incapacitating the crew members and taking over the starship. At the dramatic climax of the episode, almost all of *Voyager*'s crew are lost in hallucinatory trances, transfixed by apparitions of distant spouses welcoming them home or disapproving parents sapping their confidence.

Janeway responds to the crisis much like a Victorian gothic hero-ine: she holds firmly to reason and duty, though all around her are going mad. But then the alien appears to her in the shape of her lover, Mark, whom she may never see again since her ship is stranded at the farthest corner of the known galaxy. The apparition tries to embrace her, but she pushes him away:

> *"Mark:"* What's the matter? You used to love it when I kissed you there.
>
> *Janeway:* I don't know who you are, what you are. But I won't let you touch me.
>
> *"Mark"*: What about the man on the holodeck? You didn't seem to mind him touching you, did you? In fact, I think you liked it. Now I ask you, Kath, is that fair to me? I stayed faithful to you. I vowed to wait for you no matter how long it takes. Shouldn't you do the same?
>
> *Janeway:* (*stung, and turning to him*) I haven't been unfaithful.
>
> *"Mark:* Oh, Kath . . .

She kisses him and enters a catatonic trance.

The story of the rational and courageous Captain Janeway seduced and undone by a simulated kiss reflects a common anxiety about the new technologies of simulation. Do we believe that kissing a holo-gram (or engaging in cybersex) is an act of infidelity to a flesh-and-blood partner? If we could someday make holographic adventures as compelling as *Lucy Davenport*, would the power of such a vividly real-ized fantasy world destroy our grip on the actual world? Will the in-creasingly alluring narratives spun out for us by the new digital technologies be as benign and responsible as a nineteenth-century novel or as dangerous and debilitating as a hallucinogenic drug?

Alien Kisses

The paralyzing alien kiss is the latest embodiment of the fear with which we have greeted every powerful new representational technology—from the bardic lyre, to the printing press, to the secular theater, to the movie camera, to the television screen. We hear versions of the same terror in the biblical injunction against worshiping graven images; in the Homeric depiction of the alluring Sirens' songs, drawing sailors to their death; and in Plato's banishing of the poet from his republic because "he stimulates and strengthens an element which threatens to undermine the reason" with his fraudulent "phantasms."[6] All the representational arts can be considered dangerously delusional, and the more entrancing they are, the more disturbing. The powerful new storytelling technologies of the twentieth century have brought on an intensification of these fears. While the *Star Trek* writers imagine holodeck versions of *Beowulf* and *Jane Eyre*, a widely read and influential dystopian tradition has depicted such futuristic entertainment forms as intrinsically degrading.

Aldous Huxley's *Brave New World* (1932), set six hundred years from now, describes a society that science has dehumanized by eliminating love, parenthood, and the family in favor of genetic engineering, test-tube delivery, and state indoctrination. Books are banned, and science has come up with a substitute form of storytelling to delight the masses. In one of the novel's most memorable scenes the unspoiled hero, called the Savage (since he grew up with a biological mother in a far-off American Indian village), goes on a date to the "feelies" with Lenina, a satisfied child of the state. They are seated in the popular Alhambra theater, which is a kind of high-tech version of the plush movie palaces of the 1930s:

> Sunk in their pneumatic stalls, Lenina and the Savage sniffed and listened . . .
> The house lights went down; fiery letters stood out solid and as

though self-supported in the darkness. THREE WEEKS IN A HELICOPTER. AN ALL-SUPER-SINGING, SYNTHETIC-TALKING, COLOURED, STEREOSCOPIC FEELY. WITH SYNCHRONIZED SCENT-ORGAN ACCOMPANIMENT.

"Take hold of those metal knobs on the arms of your chair," whispered Lenina. "Otherwise you won't get any of the feely effects." (P. 134)

The attraction of the feely is an extension of the attraction of the movie and the talkie. The exuberant musicals of the early sound era are parodied by Huxley's description of the feely's foolish plot, which relies on arresting helicopter views, lots of sex, and characters who are constantly bursting into song. Writing in the age of the Hollywood star, Huxley describes the feely actors as simultaneously larger than life and less than human: a "gigantic Negro" and "a golden-haired young brachycephalic Beta-Plus female" who look "dazzling and incomparably more solid-looking than they would have seemed in actual flesh and blood, far more real than reality" (p. 134). When these too-real characters kiss, the Savage experiences for the first time the wonders of erotic engineering:

> The Savage started. That sensation on his lips! He lifted a hand to his mouth; the titillation ceased; let his hand fall back on the metal knob; it began again. . . . "Ooh-ah! Ooh-ah!" the stereoscopic lips came together again, and once more the facial erogenous zones of the six thousand spectators in the Alhambra tingled with almost intolerable galvanic pleasure. "Ooh . . ." (Pp. 134–35)

After the movie, the Savage feels debased by his own arousal. He rejects the eager Lenina and goes home instead to read *Othello*.

The horror of the feely theater lies in knowing that your intense responses have been calculated and engineered, in knowing that a technician has set the male voice at "less than 32 vibrations per second" to achieve an automatic erotic effect and has reduced the lips of all the individual audience members to just so many "facial erogenous zones" to be stimulated by galvanic means, like so many light bulbs to be flipped on.

Ray Bradbury offered a remake of the same media nightmare at the beginning of the television era. In *Fahrenheit 451* (1953), a future dictatorship keeps the populace amused and docile with raucous "televisors," sound and image systems embedded in living room walls at great expense and dedicated to incoherent but arresting entertainment. Televisor parlors are primitive holodecks in which housewife viewers converse with on-screen characters by reading from scripts in answer to their cues. Bradbury's hero, Montag (named after a paper company), is a "fireman" whose job is burning books. The novel charts his awakening from destroyer to preserver of book culture. Montag's wife, who has forgotten all the actual events of her life, has pressured him into buying three televisor walls and is pleading for the fourth so that she can be with her "parlor families" all day. In one key scene, Montag observes his wife and her friends sitting in rapt enjoyment of the disturbingly nonlinear televisor presentations:

> On one wall a woman smiled and drank orange juice simultaneously. How does she do both at once? thought Montag, insanely. In the other walls an x-ray of the same woman revealed the contracting journey of the refreshing beverage on its way to her delighted stomach! Abruptly the room took off on a rocket flight into the clouds, it plunged into the lime-green sea where blue fish ate red and yellow fish. A minute later, Three White Cartoon Clowns chopped off each other's limbs to the accompaniment of immense incoming tides of laughter. Two minutes more and the room whipped out of town to the jet cars wildly circling an arena, bashing and backing up and bashing each other again. Montag saw a number of bodies fly in the air. (Pp. 93–94)

As the housewives exclaim with delight at the entertainment, Montag pulls the switch, causing the images to drain away "as if the water had been let from a gigantic crystal bowl of hysterical fish." But

the damage remains, for when Montag tries to engage them in conversation about the coming war, the women cannot take in the reality of the situation. "It's always someone else's husband dies," they
agree, fidgeting anxiously before the now empty walls (p. 94). Like
Janeway and her crew in the grip of the alien hallucination, the televisor viewers are mesmerized by an illusion so intense that it blocks
out imminent danger.

The housewives' psychological and moral paralysis is a direct consequence of the virtues of the technology, namely, its power to appeal
to the senses of vision and hearing with stunning immediacy. In the
words of Montag's mentor, Faber (named for the pencil), the televisors are evil because they create "an environment as real as the
world." Books are praised as a better representational technology by
virtue of their limitations; their meager sensory input makes their illusions easier to resist. "You can shut them and say, 'Hold on a moment' " (p. 84). But with the new multisensory media, the populace is
overpowered.

For Huxley and Bradbury, the more persuasive the medium, the
more dangerous it is. As soon as we open ourselves to these illusory
environments that are "as real as the world" or even "more real than
reality," we surrender our reason and join with the undifferentiated
masses, slavishly wiring ourselves into the stimulation machine at the
cost of our very humanity. In this dystopian view, the new entertainment technologies are a means of stripping away the language and
culture that give life meaning and of reducing us to a state of abject
bestiality. When the Savage complains that he prefers the works
of Shakespeare because the feelies "don't mean anything," the
spokesman for the technostate assures him that "they mean a lot of
agreeable sensations for the audience" (p. 391). Why would the
docile populace want a narrative art form that helps them to better
understand themselves when they can enjoy a love scene on a sensu-
ous bearskin rug whose "every hair . . . [can] be separately and dis-
tinctly felt"?

Starting in the 1970s and 1980s, the same fears provoked by the advent of film and television began to be expressed against videogames, which added interactivity to the sensory allures of sight, sound, and motion. Critics have condemned the too-easy stimulation of electronic games as a threat to the more reflective delights of print culture. A prominent film critic, for instance, recently lamented the fact that his sons have deserted Dickens for shoot-'em-up computer games, which "offer a kind of narrative, but one that yields without resistance to the child's desire for instant gratification."[7] In recent dystopian literature, the computer screen or virtual reality helmet is as addictive and delusional as the feely or televisor. The nightmare vision of a future totalitarian state has been replaced by the equally frightening picture of a violently fragmented world organized around cyberspace, where ruthless international corporations, secret agencies, and criminal conspiracies struggle for control.

These accounts of a digital dystopia both eroticize and demonize the computer. Cyberpunk surfers are like cowboys on a new frontier or motorcycle hoodlums with a joystick in their hand instead of a motorcycle between their legs. They are outlaw pirates on an endless voyage of exploration throughout the virtual world, raiding and plundering among the invisible data hoards of the world and menaced by the stronger pirate barons who reach in and reprogram their minds. In this world, first popularized in William Gibson's *Neuromancer*, (1983) the addictive delusional experience is vividly imagined as "jacking in," that is, wiring your neurons directly into the immaterial world of "cyberspace," a word coined by the novelist to describe the virtual terrain of databanks along a surfable internet. The popular entertainment form in Gibson's gritty world is the "simstim," a way of riding in someone else's consciousness and thus experiencing the world through that person's sensorium by seeing, hearing, and feeling whatever he or she does. Case, the hero of *Neuromancer*, is addicted to the thrill of jacking in to the cyberspace data banks but is bored by the simstim as a mere "meat toy," for meat is what the body becomes when the mind finds its narcissistic love object within the machine. Yet it is hard to know which of these virtual experiences—jacking in to cyberspace or hitching a ride in a simstim—is more disturbing. In *Neuromancer* the human condition is to be faced with such choices and to flip back and forth between them with a kind of ultimate feely knob. The illusory world has become so powerfully enticing that it has subsumed physical reality itself.

But it is not just the essayists and novelists who have expressed their terrors of the emerging virtual landscape. Television shows and films have also targeted the computer as a dehumanizing representational technology. The television series *Tek War* (produced in the early 1990s by William Shatner, the actor who played the optimistic and heroic Captain Kirk on the original *Star Trek* series) is set in a future America destroyed by the illegal traffic in Tek, a hallucinogenic technology resembling a virtual reality headset. In the first episode of the series, for instance, the hero is paralyzed by powerful Tek programs, bought on the black market, that simulate his ex-wife returning to love him. When his partner arrives to tear off the helmet and bring him back to chasing bad guys, it is a scene very like the classic Western cliché of the sheriff sobering up the drunken deputy but with darker urban overtones suggesting a heroin or cocaine habit. Throughout the *Tek War* series, virtual reality technologies are explicitly equated with lethal drugs as the source of addiction, destitution, bad trips, overdose deaths, and gangster violence.

Movies have been even more lurid in their depiction of computer-based entertainment. Perhaps the most explicit filmic statement of the dangers of cyberspace is *Lawnmower Man* (1992), in which a virtual reality researcher turns a simpleminded gardener into a digital monster. In this retelling of the Frankenstein story, Dr. Larry Angelo experiments with Jobe Smith with the intention of expanding his mental abilities. Larry's first step in sending Jobe down the road to psychosis is to invite him in to play virtual reality videogames that speed up his mind, awakening neurons the rest of us leave dormant. Soon Jobe rejects books as too slow a means of learning and listens to music by jumping from one short excerpt to another. Once he has left the world of linear media behind, he quickly turns to horror-movie-style slaughter, accomplished by the sheer power of his unnaturally augmented brain. The movie climaxes with Jobe leaving his body and entering the machine, where he appears as a kind of videogame character. The virtual Jobe easily outfights the virtual image of his creator and eventually escapes into the Internet. At the very end of the movie, we hear the sound of all the telephones in the world ringing simultaneously, signaling that this superior being is on the verge of taking over the planet. In effect, the videogame will play *us* from now on. *Lawnmower Man* is the most extreme version of the dystopian vision: the representational technology as both diversion and dictator all in one.

The Thinking Woman's Feely

Which vision of digital storytelling are we to believe? Will the literature of cyberspace be continuous with the literary traditions of the *Beowulf* poet, Shakespeare, and Charlotte Brontë as the *Star Trek* producers portray it, or will it be the dehumanizing and addictive sensation machine predicted by the dystopians? Is the optimistic *Star Trek* view too pat and sentimental to be credible at all in the light of Huxleyan criticism?

We can certainly see Captain Janeway's experience as dystopian. The holodeck is in many ways exactly the sort of entertainment machine Huxley dreaded: a masterpiece of engineering aimed at inducing delusional physical sensations. No doubt the appropriate moisture and temperature of Lord Burleigh's kiss have been as carefully calculated as the sensations produced by the feely knob. But unlike the helpless fantasy addicts of the dystopian stories, Janeway is the master of the apparatus that is creating the illusion. This is made clear when she returns to the holodeck after her initial hallucinations to check it for a malfunction and is eagerly greeted by her virtual lover:

> Burleigh: Lucy, thank God you've come back. (*Notices her uniform*) But why are you dressed so strangely?

Janeway: It's a costume.

Burleigh: You'd look lovely in anything. (*Takes her hand*) I've thought of you constantly, remembered your touch, your perfume, your lips.

Janeway: (*Eyes closed, as if surrendering to his magnetism*) Computer, delete character!

Even as she swoons in an embrace, Janeway is in control of the mirage. In Bradbury's terms, she can shut the book. Lord Burleigh is deliciously enticing but unenslaving, just as movie and television heart throbs from Clark Gable's Rhett Butler to George Clooney's Dr. Doug Ross have proved to be, despite dystopian fears.

The *Star Trek* story can be seen as a fable differentiating humane and meaningful digital storytelling from the dehumanizing illusions that the dystopians warn about. Janeway is paralyzed by her hallucination of her lover, Mark, because it is too literal a transcription of her fantasies. The alien treats human consciousness like a stimulus–response machine. The holonovel, on the other hand, is aimed not at Janeway's neurons but at her imagination. Although it offers the pleasures of an art form "more real than reality," it is clearly make-believe. At the end of the episode, Janeway is skipping her regular visit to the holodeck to think about the issues the enemy hallucinations have raised. Now that the alien is defeated by the superior telepathic powers of another female crew member, Janeway thinks, "In a way, maybe he did us all a favor. Maybe it's better to look those feelings in the eye than to keep them locked up inside." The holodeck, like any literary experience, is potentially valuable in exactly this way. It provides a safe space in which to confront disturbing feelings we would otherwise suppress; it allows us to recognize our most threatening fantasies without becoming paralyzed by them. Like a magical starship designed for safely exploring the distant quadrants of the galaxy, the holodeck is an optimistic technology for exploring inner life. For Captain Janeway, a person of Victorian integrity, such an exploration brings the benefit of self-knowledge. It is not paralyzing. It sends her back to the real world all the stronger.

The holonovel offers a model of an art form that is based on the most powerful technology of sensory illusion imaginable but is nevertheless continuous with the larger human tradition of storytelling, stretching from the heroic bards through the nineteenth-century novelists. The feely (and its successors) offers an opposing image of a sensation-based storytelling medium that is intrinsically degrading, fragmenting, and destructive of meaning, a medium whose success implies the death of the great traditions of humanism, or even a fundamental shift in human nature itself. Neither vision of the future refutes the other. Together they sum up the hopes and fears aroused by the increasingly visceral representational technologies of the twentieth century. As these utopian and dystopian fictions remind us, we rely on works of fiction, in any medium, to help us understand the world and what it means to be human. Eventually all successful storytelling technologies become "transparent": we lose consciousness of the medium and see neither print nor film but only the power of the story itself. If digital art reaches the same level of expressiveness as these older media, we will no longer concern ourselves with how we are receiving the information. We will only think about what truth it has told us about our lives.

Chapter 2

Harbingers of the Holodeck

The final quarter of the twentieth century marks the beginning of the digital age. Starting in the 1970s, computers have become cheaper, faster, more capacious, and more connected to one another at exponential rates of improvement, merging previously disparate technologies of communication and representation into a single medium. The networked computer acts like a telephone in offering one-to-one real-time communication, like a television in broadcasting moving pictures, like an auditorium in bringing groups together for lectures and discussion, like a library in offering vast amounts of textual information for reference, like a museum in its ordered presentation of visual information, like a billboard, a radio, a gameboard, and even like a manuscript in its revival of scrolling text. All the major representational formats of the previous five thousand years of human history have now been translated into digital form. There is nothing that human beings have created that cannot be represented in this protean environment, from the cave paintings of Lascaux to real-time photographs of Jupiter, from the Dead Sea Scrolls to Shakespeare's First Folio, from walk-through models of Greek temples to Edison's first movies. And the digital domain is assimilating greater powers of representation all the time, as researchers try to build within it a virtual reality that is as deep and rich as reality itself.

The technical and economic cultivation of this fertile new medium of communication has led to several new varieties of narrative entertainment. These new storytelling formats vary from the shoot-'em-up videogame and the virtual dungeons of Internet role-playing games to the postmodern literary hypertext. This wide range of narrative art holds the promise of a new medium of expression that is as varied as the printed book or the moving picture. Yet it would be a mistake to compare the first fruits of a new medium too directly with the accustomed yield of older media. We cannot use the English theater of the Renaissance or the novel of the nineteenth century or even the average Hollywood film or television drama of the 1990s as the standard by which to judge work in a medium that is going through such rapid technical change.

In 1455, Gutenberg invented the printing press—but not the book as we know it. Books printed before 1501 are called incunabula; the word is derived from the Latin for swaddling clothes and is used to indicate that these books are the work of a technology still in its infancy. It took fifty years of experimentation and more to establish such conventions as legible typefaces and proof sheet corrections; page numbering and paragraphing; and title pages, prefaces, and chapter divisions, which together made the published book a coherent means of communication. The garish videogames and tangled Web sites of the current digital environment are part of a similar period of technical evolution, part of a similar struggle for the conventions of coherent communication.[1]

Similarly, new narrative traditions do not arise out of the blue. A particular technology of communication—the printing press, the movie camera, the radio—may startle us when it first arrives on the scene, but the traditions of storytelling are continuous and feed into one another both in content and in form. The first published books were taken from the manuscript tradition. Malory's *Morte d'Arthur,* written in manuscript in 1470, drew on prose and poetry versions of the Camelot legend in both French and English, which in turn drew on centuries of oral storytelling. The elements of the story were all there already: the rise and fall of the hero Arthur, the gallantry of the knights, the love between Guinevere and Lancelot, and the destruction of the Round Table through civil war. But Malory's prose brought these elements together and introduced colloquial dialogue, more consistent plotting, and a pervasive tone of nostalgia. Fifteen years later, William Caxton took Malory's separate tales and bound them together into a single volume, with descriptive chapter headings that lured readers into the story. Only then, after such long episodic narratives were commonplace in publishing, could Cervantes write a contemporary tale like *Don Quixote* (1605), which marks the beginning of the European novel.

We can see the same continuities in the tradition that runs from nineteenth-century novels to contemporary movies. Decades before the invention of the motion picture camera, the prose fiction of the nineteenth century began to experiment with filmic techniques. We can catch glimpses of the coming cinema in Emily Brontë's complex use of flashback, in Dickens' crosscuts between intersecting stories, and in Tolstoy's battlefield panoramas that dissolve into close-up vignettes of a single soldier. Though still bound to the printed page, storytellers were already striving toward juxtapositions that were easier to manage with images than with words.

Now, in the incunabular days of the narrative computer, we can see how twentieth-century novels, films, and plays have been steadily pushing against the boundaries of linear storytelling. We therefore have to start our survey of the harbingers of the holodeck with a look at multiform stories, that is, linear narratives straining against the boundary of predigital media like a two-dimensional picture trying to burst out of its frame.

The Multiform Story

I am using the term *multiform story* to describe a written or dramatic narrative that presents a single situation or plotline in multiple versions, versions that would be mutually exclusive in our ordinary experience. Perhaps the best-known example of a multiform plot is Frank Capra's beloved Christmas story, *It's a Wonderful Life* (1946), in which hardworking, benevolent George Bailey, as played by Jimmy Stewart, is given a vision of what his town would have been like if he had never lived. The film juxtaposes two divergent pictures of George's hometown: the present-time Bedford Falls, in which George has saved his father's small savings and loan bank, married the town librarian, and been a benefit to his community, and a town originally called Bedford Falls but renamed Pottersville by the evil big banker Potter, a town in which there is no savings and loan to offer mortgages, the librarian is a bitter old maid, and everyone's life is poorer and meaner without George's compassionate guidance. The movie as a whole pivots around the moment when George, facing ruin and remembering all the disappointments of his life, is standing on a bridge contemplating suicide. The whimsical angel Clarence persuades him to live by running a kind of simulation experiment—a replay of the past thirty years in Bedford Falls as it would have turned out if George had never been born. In this film the multiform story format works as a kind of scientific proof of the meaning of one person's life.

But for many postmodern writers, the quintessential multiform narrative is the much darker story in Jorge Luis Borges's "The Garden of Forking Paths" (1941). Here the pivotal moment is a seemingly meaningless act of murder. The narrator, Dr. Yu Tsun, is a German spy during World War I who knows that he is on the verge of capture. He resolves to murder a man named Steven Albert, whose name he has selected from the phone book. Albert, by coincidence, has devoted his life to studying an incoherent novel (which is also called *The Garden of Forking Paths*) written by Ts'ui Pên, an an-

cestor of the narrator. As Albert explains to Yu Tsun, the story of the
forking path is really a labyrinth because it is based on a radical
reconception of time:

> In all fiction, when a man is faced with alternatives he chooses one at
> the expense of the others. In the almost unfathomable Ts'ui Pên, he
> chooses—simultaneously—all of them. He thus *creates* various fu-
> tures, various times which start others that will in their turn branch
> out and bifurcate in other times. (P. 98)

Time in Ts'ui Pên's world is not an "absolute and uniform" line but
an infinite "web" that "embraces every possibility." Albert tells his future murderer that they are living in a world of similarly bifurcating
time, full of many alternate realities:

> We do not exist in most of them. In some you exist and not I, while in
> others I do, and you do not, and in yet others both of us exist. In this
> one, in which chance has favored me, you have come to my gate. In
> another, you, crossing the garden, have found me dead. In yet
> another, I say these very same words, but am an error, a phantom.
> (P. 100)

As Yu Tsun gets closer to committing the murder, he is aware of a
"pullulation," a splitting of reality. Like the characters in Ts'ui Pên's
story, he is choosing multiple alternatives, creating various futures simultaneously:

> It seemed to me that the dew-damp garden surrounding the house
> was infinitely saturated with invisible people. All were Albert and
> myself, secretive, busy and multiform in other dimensions of time.
> (Pp. 100–101)

The notion of multiple possible worlds seems at first to absolve the
narrator of moral responsibility and to make the deed much easier.
He murders the unsuspecting Albert while his back is turned, choosing his moment in order to be as merciful as possible. It is a dispas-
sionate crime, a triumph of cryptography. Yu Tsun has succeeded in

sending a message alerting the Germans to attack a city named Albert by causing his own name to appear linked with the name of his victim in newspapers. Since Yu Tsun does not believe in the German cause, the murder is a deeply meaningless act of pure communication. Yet the story ends with the narrator full of "infinite penitence and sickness of heart" (p. 101). The fact that Yu Tsun's experience of life is only a slender thread in the infinite web of his possible lives does not change the fact that he is firmly embedded in his single lived reality.

A similarly pullulating moment underlies Delmore Schwartz's chilling story "In Dreams Begin Responsibilities," first published in 1937. The story is told by a 21-year-old narrator who is dreaming that he is watching a silent movie of the day his father proposed to his mother on a date at Coney Island. His parents are engagingly vulnerable and hopeful, though it is achingly clear that they will make one another quite miserable. In the central scene of the story, the narrator watches as his father confidently orders an ocean-view table in the best restaurant on the boardwalk and awkwardly makes his proposal; his mother weeps with joy as she accepts. At this moment the narrator rises from his seat in the theater and begins to shout at the characters on the screen: "Don't do it. It's not too late to change your minds, both of you. Nothing good will come of it, only remorse, hatred, scandal, and two children whose characters are monstrous" (p. 6). But the usher forces him to sit down while the unchangeable past continues to unfold on the screen .

Near the end of the story, the narrator's mother feels compelled to enter a palmistry booth. His father grudgingly waits around with her until the fortune-teller appears.

> But suddenly my father feels that the whole thing is intolerable; he tugs at my mother's arm, but my mother refuses to budge. And then, in terrible anger, my father lets go of my mother's arm and strides out, leaving my mother stunned. She moves to go after my father, but the fortune-teller holds her arm tightly and begs her not to do so, and I in

my seat am shocked more than can ever be said, for I feel as if I were walking a tight-rope a hundred feet over a circus-audience and suddenly the rope is showing signs of breaking, and I get up from my seat and begin to shout once more the first words I can think of to communicate my terrible fear . . . and I keep shouting: "What are they doing? Don't they know what they are doing? Why doesn't my mother go after my father? If she does not do that, what will she do? Doesn't my father know what he is doing?" (P. 8)

As these alternate futures pullulate around his mother in the fortune-teller's booth, the dreamer is scolded by the usher in words that sum up his feelings of panic. "You can't carry on like this," he is told. "Everything you do matters too much" (pp. 8–9). The danger for the narrator is the same one George Bailey faces: the danger of wishing never to have been born and having your wish come true. The story ends here as he wakes up "into the bleak winter morning" of his twenty-first birthday, into the reality that is the result of his mother's moment of choice.

Schwartz's story was arresting when it came out, as Irving Howe remembers, in its depiction of the inexorability of the past as a movie reel that "must run its course; it cannot be cut; it cannot be edited."[2] But from the perspective of the 1990s, we can see that the originality of the story also lies in its dramatization of the narrator's position in the audience as he attempts to turn a linear, passive medium into an interactive one. The question that is tormenting him is not whether he can bear to witness the past by watching the painful film unroll, but whether he would choose to change it if he could. Would the dreamer redream his parents' unhappy love story knowing that if he did so he might never wake up? The multiform story is an expression of the anxiety aroused by posing such choices to oneself.

To explore such questions concretely in linear media, we usually have to enter the realm of science fiction. In fact, Schwartz's narrator's disturbing fantasy of undoing his parents' marriage by interrupting their moment of betrothal is replayed as a farcical adventure in the Robert Zemeckis hit film *Back to the Future* (1985). When the hero, teenager Marty McFly, time-travels back to the 1950s, his photograph of himself and his siblings starts to fade as his bumbling actions make his parents' marriage less and less likely. To survive his adventure, Marty must make sure his parents kiss at a particular moment of the upcoming high school dance, and he is appalled to realize just how unlikely their union seems to be. The moment of the kiss is so pivotal that it is repeated in the sequel movie, with a *second* time-traveling Marty seeing it (and risking its disruption) all over again. As George McFly stands on the dance floor in the school gym, unable to work up the courage to embrace his very willing partner, Marty No. 1, who has been playing guitar onstage to keep the mood going, starts to fade out of existence, a victim of his father's sexual cowardice. In the sequel version, Marty No. 2 is suspended on the catwalk over the stage, fleeing the villain and in danger of falling, much like Schwartz's narrator, who feels suspended on a mental "tightrope" as his mother stands between the fortune-teller and her fate.

Of course, in the Hollywood version of the disrupted proposal story there is a much happier ending: not only do Marty's father and mother get together, but George McFly, who would otherwise remain an ineffectual and cowardly nerd, rescripts his life when he makes a fist and hits the evil bully, Biff. Marty returns to a world in which his father is a successful science fiction writer, his mother is thin and cheerful, his sister is popular, his brother has a good job, and he has unrestricted access to the family car. He has achieved a familiar twentieth-century adolescent fantasy: to totally remake his family according to his own desires.

Part of the impetus behind the growth of the multiform story is the dizzying physics of the twentieth century, which has told us that our common perceptions of time and space are not the absolute truths we had been assuming them to be. The emotional conundrums of the Einsteinian view have been most explicitly explored in Alan Lightman's *Einstein's Dreams* (1993), which offers poetic vignettes of human life as it might be under other systems of time. For instance, in a world in which "time has three dimensions, like space," a man stands on a balcony in Berne thinking about a woman in Fribourg. "His hands grip the metal balustrade, let go, grip again. Should he visit her. Should he visit her?" (pp. 18–19). In one world he decides not to go and instead "keeps to the company of men" until three years later he meets a nice woman in a clothing shop in Neuchâtel who eventually comes to live with him and with whom he contentedly grows old. In another he decides he "must see" the woman in Fribourg despite her volatility; he leaves his job and moves to Fribourg, where they live stormily together and "he is happy with his anguish." In the third world he is also driven to see her but they merely talk for an hour and then she says she must leave; he returns to his balcony feeling empty. How do people live in a world where they are conscious of the world splitting in three at every decision point, a world in which there are infinite alternatives to every situation? Lightman imagines it this way:

> Some make light of decisions, arguing that all possible decisions will occur. In such a world, how could one be responsible for his actions? Others hold that each decision must be considered and committed to, that without commitment there is chaos. Such people are content to live in contradictory worlds, so long as they know the reason for each. (P. 22)

Lightman's story, like Borges's, is a haunting evocation of the world of ordinary experience, of our own perception of moments of choice that teem with multiple possibilities, all of which seem authentic—if not in their "quantum signatures" (as science fiction writers would say), then in their emotional signatures. We know what it feels like to stand on that balcony and consider three possible lives that all feel real. We are outgrowing the traditional ways of formulating this experience because they are not detailed or comprehensive enough to express our sense of the pullulating possibilities of life.

The most successful attempt to portray multiple alternate realities within a coherent linear story is Harold Ramis's farcical movie

Groundhog Day (1993), in which a selfish and bitter weatherman named Phil is forced to relive a single winter's day in a hick Pennsylvania town until he gets it right. The film works in part because it never attempts to explain why Phil keeps waking up on the same day.[3] It just puts him in this absurd situation and watches what he does about it. The day is detailed as a series of witty variations on a set of comic motifs. Rushing to do a broadcast about the appearance of the groundhog, Phil is accosted by an overfriendly high school friend, Ned, who tries to sell him insurance. In his haste to get away from the irritating Ned, Phil steps off a curb and into a deep puddle of water. The scene is shown four times with interesting variations, including one in which Phil embraces Ned first and with so much intensity that Ned is the one to run away. The pleasure for the audience is in savoring the variations, wondering how Phil is going to play it this time. Phil's life is not an inexorable film reel, like the Coney Island date in Schwartz's dream-movie, but an endless series of retakes. When he sets out to seduce his producer, Rita, he repeats his date with her endlessly, revising every aspect of it to suit her tastes and fantasies, only to wind up slapped and rejected many times over. Eventually Phil learns to live his one day as a better person; he takes up the piano, prevents the accidents he knows are due to happen, and opens his heart to the people he formerly looked on with contempt. Once he gets the day right, he wins Rita's love and finally wakes up on February 3.

Groundhog Day is, in its way, an updating of the familiar marriage plot, like the ones in Jane Austen's novels, in which courtship is depicted as a process of moral education. Because Phil is a man of the 1980s, his learning is conducted in the form of an educational simulation—the opposite of the one the angel Clarence runs for George Bailey—in which the town is held constant and only the protagonist changes. Because of his simulation structure, *Groundhog Day*, though it has none of the shoot-'em-up content of videogames, is as much like a videogame as a linear film can be.

Multiform stories often reflect different points of view of the same event. The classic example of this genre is *Rashomon* (1950), the Kurosawa film in which the same crime is narrated by four different people: a rape victim; her husband, who is murdered; the bandit who attacks them; and a bystander. The increasing moral confusion of their accounts in part reflects the postwar cultural crisis in Japan. Similarly, in Milorad Pavic's *Dictionary of the Khazars* (1988) the impending dissolution of Yugoslavia is prefigured by the fragmentary account of a mythical lost tribe whose history is known through conflicting Christian, Jewish, and Moslem versions. The book is designed as three incomplete "dictionaries" (really more like encyclopedias), which represent the three religious traditions and have conflicting entries for the same events. Although published in a bound volume, the book is not meant to be read in consecutive order, as the author tells the reader:

> The three books of this dictionary . . . can be read in any order the reader desires; he may start with the book that falls open as he picks up the dictionary . . . *The Khazar Dictionary* can also be read diagonally, to get a cross-section of all three registers—the Islamic, the Christian, and the Hebrew . . . He may move through the book as through a forest from one marker to the next . . . He can rearrange it in an infinite number of ways, like a Rubik cube . . . Each reader will put together the book for himself, as in a game of dominoes or cards, and as with a mirror, he will get out of this dictionary as much as he puts into it. (Pp. 12–13)

The fragmentation of the story structure represents patterns of historical fragmentation, and the patterns of readings echo the characters' efforts to reconstruct the past in order to restore a lost coherence.

As this wide variety of multiform stories makes clear, print and motion picture stories are pushing past linear formats not out of mere playfulness but in an effort to give expression to the characteristically twentieth-century perception of life as composed of parallel possibilities. Multiform narrative attempts to give a simultaneous form to these possibilities, to allow us to hold in our minds at the same time multiple contradictory alternatives. Whether multiform narrative is a reflection of post-Einsteinian physics or of a secular society haunted by the chanciness of life or of a new sophistication in narrative thinking, its alternate versions of reality are now part of the way we think, part of the way we experience the world. To be alive in the twentieth century is to be aware of the alternative possible selves, of alternative possible worlds, and of the limitless intersecting stories of the actual world. To capture such a constantly bifurcating plotline, however, one would need more than a thick labyrinthine novel or a sequence of films. To truly capture such cascading permutations, one would need a computer.

The Active Audience

When the writer expands the story to include multiple possibilities, the reader assumes a more active role. Contemporary stories, in high and low culture, keep reminding us of the storyteller and inviting us to second-guess the choices he or she has made. This can be unsettling to the reader, but it can also be experienced as an invitation to join in the creative process.

Italo Calvino's *If on a Winter's Night a Traveler* (1979) is a novel in the form of a long meditation on fiction making, a story that keeps unraveling and restarting itself. In a world that is perceived as a vast interconnected web, how is the author to know which thread to pull on first? How can he hope "to establish the exact moment in which a story begins"?

> Everything has already begun before, the first line of the first page of every novel refers to something that has already happened outside the book. . . . The lives of individuals of the human race form a constant plot, in which every attempt to isolate one piece of living that has a meaning separate from the rest—for example, the meeting of two people, which will become decisive for both—must bear in mind that

each of the two brings with himself a texture of event, environments, other people, and that from the meeting, in turn, other stories will be derived which will break off from their common story. (P. 153)

The beginning of any story is fraught with possibilities:

> On the wall facing my desk hangs a poster somebody gave me. The dog Snoopy is sitting at a typewriter, and in the cartoon you read the sentence, "It was a dark and stormy night. . . ." Every time I sit down here I read, "It was a dark and stormy night . . ." and the impersonality of that *incipit* seems to open the passage from one world to the other, from the time and space of here and now to the time and space of the written word; I feel the thrill of a beginning that can be followed by multiple developments, inexhaustibly. (Pp. 176–77)

The commitment to any particular story is a painful diminution of the intoxicating possibilities of the blank page. Calvino's fiction is offering a new kind of story pleasure, a delight not in the tale but in the fertile mind of the writer.

It is not just intellectual fiction that has become so self-aware. Evidence of the same tendency in popular fiction is as close at hand as two of my son's recent Christmas presents. Popular comic book writer Mike Baron introduces a collection of the first five *Nexus* issues with a chatty description of his collaboration with his graphic artist partner, Steve Rude. He shares with the readers his perspective on one of the main villains of the ongoing story: "I think Nexus' universe would be a duller place without Ursula, but the Dude is constantly howling for her blood. I've saved her life several times in impassioned late-night phone calls." When the writer talks about her in this way, Ursula loses credibility as a fictional character but she becomes more interesting as an aspect of her creators' imagination. The important contest for the reader, the focus of dramatic suspense, is not the one between Nexus and Ursula but between Baron and Rude.

Giving the audience access to the raw materials of creation runs the risk of undermining the narrative experience. I lose patience with

Calvino when he repeatedly dissolves the illusion. When in *Groundhog Day* the conversation at a bar between Phil and Rita is repeated over and over again to show how Phil changes his pickup routine over several days, the sequence looks confusingly like a series of retakes of a single movie scene; I am reminded that I am watching Bill Murray and Andie MacDowell repeating lines for the camera. Nevertheless, calling attention to the process of creation in this way can also enhance the narrative involvement by inviting readers/viewers to imagine themselves in the place of the creator.

Murder mysteries, for example, count on the reader to be aware of the conventions of the form and to anticipate multiple arrangements of the elements provided by the author. Is that odd-looking woman outside the murder scene an important witness? A murderer? The next victim? Is she perhaps not a woman at all but a man in disguise? Serial narratives like Victorian novels or contemporary television shows also sustain audience involvement between installments by skillfully setting up plot patterns that encourage speculation on which possibilities will be developed. Comic book franchises acknowledge and encourage the audience's free-form fantasies by publishing special series devoted to events that are contrary to the official history of the characters but full of interesting narrative possibilities. Marvel Comics uses its *"What If . . . ?"* monthly series to explore such questions as "What if Spiderman's uncle had not died?" and "What if Spiderman had never gotten superpowers?"; and DC Comics uses its forty-eight-page *Elseworlds* issues (twice the size of the usual monthly) to imagine Superman transported to the Metropolis of Fritz Lang's 1926 film or Batman born into Victorian England and fighting Jack the Ripper. These efforts assume a sophistication on the part of the audience, an eagerness to transpose and reassemble the separate elements of a story and an ability to keep in mind multiple alternate versions of the same fictional world.

Although television viewers have long been accused of being less actively engaged than readers or theatergoers, research on fan culture provides considerable evidence that viewers actively appropriate the stories of their favorites series.[4] Fan culture has grown over the past decades through conventions, underground magazines, and the trading of home videos. The Internet has accelerated this growth by providing a medium in which fans can carry on (typed) conversations with one another and often with the producers, writers, and stars of ongoing series. Much of this discourse is focused on the consistency of the shows, with careful debate on such issues as whether a supporting character on a sitcom is a widower or a divorcé or which fictional New York City cop most deserves a promotion.

In addition to sharing critical commentary and gossip, fans create their own stories by taking characters and situations from the series and developing them in ways closer to their own concerns. *Star Trek* fans in particular have produced a vast literature of alternate adventures over the thirty years since the original series aired. Women writers have created stories in which the female characters take over the ship or refuse the advances of the notoriously lecherous Captain Kirk. The romantic rivalry of the aggressive Worf and the egotistical Riker for the voluptuous Deanna Troi has inspired many more fanwritten stories than episodes of the *Next Generation* series in which it was introduced. With the advent of the VCR, a new branch of fan literature has arisen in which actual scenes from the broadcast programs are reedited into new stories. Kirk and Spock, whose friendship is a centerpiece of the original series, have been reinterpreted as lovers through the magic of videotape. This "textual poaching," as media critic Henry Jenkins has called it, has become even more widespread on the World Wide Web, which functions as a global fanzine. Although some copyright holders have protested, fans have little trouble obtaining digital images and even digital video clips from their favorite series, which they put to their own use on personal Web pages. The imaginative involvement of fans gives them a strong sense of entitlement to the images associated with their favorite shows. When the Microsoft Network closed off its official *Star Trek* Web site, "*Star Trek* Continuum," to users with non-Microsoft Web browsers, fans organized a protest campaign and enjoyed pointing out how superior their own Web pages were to the official site.

The most active form of audience engagement comes in role-playing clubs. Fans of fantasy literature from Tolkien to space operas have joined together for live-action role-playing (LARP) games in which they assume the roles of characters in the original stories to make up new characters within the same fictional universe. This youthful gaming world, which began with twelve-year-olds playing *Dungeons and Dragons* in the 1970s, has grown by the 1990s to include long-standing, organized role-playing groups composed of dozens of college students and young professionals.[5] Some of these games, like a San Francisco vampire group of post-college-age players, last for several years, with players maintaining the same character over the course of the game. Others, like many of those created for the Assassins' Guild, a role-playing club at MIT, can be over in an intense weekend. Some of the games focus on jousts and ambushes, others on elaborate political negotiation, and still others on skillful improvisations of dramatic scenes. In all of them, the players share a sense of exploring a common fictional landscape and inventing their stories as they go along.

Role-playing games are theatrical in a nontraditional but thrilling way. Players are both actors and audience for one another, and the events they portray often have the immediacy of personal experience. For instance, in a live-action game at MIT set in a world populated by characters based on Shakespeare's plays, Seth McGinnis, a graduating senior, had the secret identity of Puck from *Midsummer Night's Dream*. Puck was disguised from the other players as a member of a troupe of traveling actors who stage a performance of the Pyramus and Thisbe scene from *Midsummer Night's Dream* with Puck playing the role of the lover Pyramus. Seth decided to take advantage of the confusion that occurs as everyone leaves the "theater" to use his fairy powers to create an illusory wall between a prisoner and his guards, thus allowing the prisoner to escape. Puck's wall actually consisted of one of the game masters standing for five minutes with arms stretched across the entrance to a stairway leading from the MIT classroom designated as a town square to the MIT classroom designated as the tavern. Pyramus and Thisbe talk to one another through a similar illusory wall, portrayed by a comically clumsy actor who uses his fingers to make a chink through which the lovers whisper. The crudely portrayed wall is an enduringly charming bit of stage business within the original play and a gentle reminder of the make-believe of theater itself. The wall in the game, like the wall in the play, was a consensual reality. The players joined in the creation of the illusion by poking at the wall, expressing amazement at its sudden appearance, and proclaiming that they could not see around it. But unlike actors in a play, the players were also genuinely puzzled about how the wall was created and by whom. Puck's wall had the arresting presence of a spontaneous event. It will not last as long as Shakespeare's, but for the people playing the game that night it was even more dramatically compelling.

Live theater has been incorporating the same qualities of spontaneity and audience involvement for some time. Improvisational groups solicit suggestions from audience members and offer them the pleasure of performance combined with the pleasure of witnessing creative invention. Participatory dinner theater casts the members of the audience as bit players in a group event, such as a comic wedding, jury trial, or wake. Mainstream audiences have recently accepted being addressed from the stage as schoolchildren or PTA members, and have even followed actors around a New York townhouse.[6] Commercial role-playing games mix actors with paying guests who solve a mystery or enact a spy drama over a weekend at a vacation resort. In all of these gatherings, the attraction lies in inviting the audience onto the stage, into the realm of illusion. These are all holodeck experiences without the machinery.

And the machinery—all but three-dimensional holograms— seems not that far away. Since the 1980s, gaming environments called MUDs (Multi-User Domains) have allowed distant players on the Internet to share a common virtual space in which they can "chat" with one another (by typing) in real time.[7] Words typed by fellow players all over the planet appear on each player's screen as the players improvise scenes together and collectively imagine fictional worlds. As the social psychologist Sherry Turkle has persuasively demonstrated, MUDs are intensely "evocative" environments for fantasy play that allow people to create and sustain elaborate fictional personas over long periods of time. Every day, and particularly every night, thousands of people forsake real life (RL) and meet in virtual space "in character" (IC) to play out stories based on favorite books, movies, or television shows. This new kind of adult narrative pleasure involves the sustained collaborative writing of stories that are mixtures of the narrated and the dramatized and that are not meant to be watched or listened to but shared by the players as an alternate reality they all live in together.

Movies in Three Dimensions

We do not have to wait for *Star Trek*'s fanciful molecular replication technology or the "emotional engineers" of *Brave New World* to see three-dimensional fictional characters standing before our eyes. The Sony IMAX Theater across from Lincoln Center in New York City is the very model of Huxley's Alhambra. Entering a lobby ringed with video screens and ticket machines, you ascend through an atrium of multistory escalators and pass through a seemingly limitless expanse of theaters until you reach, at the very top, "the BIGGEST movie screen on earth." How big is it? A video monitor is winking away over the waiting area to bombard you with the statistics. The 3-D screen is eight stories high and 100 feet wide, the size of seven elephants; the special film is ten times the size of 35mm film, is stored in a canister that is 7.5 feet in diameter, and runs in a projector that weighs 500 pounds and uses 18,000 watts of electricity. Inside you sit in a cheerful, spacious, banked theater facing the indeed enormous screen, and though there are no feely knobs, you are provided with a pair of plastic 3-D goggles with liquid crystal lenses and built-in speakers that create a "personal sound environment." The goggles are engineered so that an undetectable shutter action takes place many times a second, blanking out one eye and then the other, to send two separate images to the imaging centers of your brain. It is the combination of the slightly different left and right images that produces the appearance of three-dimensional space.

When the movie starts, the sensation is not of size or gadgetry but of a magical apparition, for the 3-D movies that are shown in this new Alhambra make conventional movies look like daguerreotypes. The world that is displayed through those lightweight and soon forgotten goggles has the depth and dimension of the actual world, where you can see around things, look left and right, and shift your focus from back to front within the same image. The size of the film means an increase in information, offering a richer and therefore more persuasive visual illusion. It is not merely a larger image but a more present reality.

For a short feature this sense of presence is exciting in itself. When I saw my first 3-D movie at Disney World's Epcot Center in the 1980s, I held my breath when a little blue bird flew out of the screen and landed right in front of my nose. I and everyone else in the audience reached out a hand to touch the bird, for we each, at our different locations, saw it right in front of us. During the viewing of a long feature, the reaching eventually subsides as the audience comes to take for granted a representational world with persuasive depth but no solidity. The question then becomes, What kinds of stories is such a high-sensory technology suited to tell us? Filmmakers have just begun to answer that question, but the first two feature films made with the IMAX technology look much more like *Star Trek*'s Lucy Davenport than like Huxley's *Three Weeks in a Helicopter*.

Across the Sea of Time (1995) is a modest story of a Russian immigrant boy, Tomas, who has magically arrived in contemporary New York to trace the path of an immigrant relative with the help of stereopticon photos from the turn of the century. The story provides a pretext for spectacular photography, including the helicopter shots

Huxley was already lamenting in the 1930s, here accompanied by the sound of violin crescendos as we swing across the Brooklyn Bridge. But these panoramas, like the billboard ads and insurance blimps caught by the camera, are there to pay the rent by making the film serve as a good tourist attraction. They are not that much more striking than the familiar two-dimensional versions or the large-format films shown in amusement parks or planetariums. The three-dimensional panoramas do become striking, however, when they are anchored by the foreground figure of the young boy. When Tomas is standing on the parapet of a skyscraper and looking at the vast spaces of the city, we are taken out of the generic landscape of tourist spectacle and placed in a very present dramatic moment. Such moments indicate that this is a technology that is ready to tell more intimate stories.

A large part of the pleasure of the film lies in the original black-and-white stereopticon photos. Even though the people in these photos appear rather like cutouts in a diorama-like scene box, the establishing of multiple planes animates them. The three-dimensional projection becomes a resurrection of the dead; we are given the ability to see them and to see the world through their eyes with stunning immediacy. The joy of a particular day on the beach at Coney Island is made palpable in the way a pair of lovers are leaning toward one another and in the weight of a girl's arm around her friend's shoulder as they laugh and enjoy their holiday. The sensation of resurrection is even stronger in a photo taken of three workers, two white and one black, digging a tunnel for a subway. We enter the deep tunnel and feel the dank, claustrophobic confinement. We look at the posture and feel the exhausting labor. Here is the very antithesis of the feely, yet it is delivered in the exact technology Huxley distrusted. These stereopticon images wedded to film are used not to distance us from reality or to present oversized, dehumanized "stars," but to bring us close to the plain working folk whose experiences make up the true but hidden history of a great city. The technology does not make them larger than life, only more present to us.

One of the reasons the subway scene works so well is that three-dimensional photography is particularly impressive for enclosed spaces. Perhaps the most successful dramatic moment comes early in the film when the boy is a stowaway on a boat leaving Ellis Island. As Tomas cowers in the cramped hull of the ship, surrounded by the cold metal of the ship's pipes and machinery, a huge but kindly-looking stranger opens the door of his hiding place, reaches forward and extends to the boy a paper lunch bag. Sitting in the audience I could almost feel the lunch bag in my lap, and I experienced the generosity of the moment almost personally because I was so physically grounded in the boy's surroundings. In a conventional movie such a moment would have to be emphasized by close-up shots of the boy's face expressing his feelings of gratitude. In a 3-D film, the audience can be so closely identified with the situation of a character that such reaction shots are unnecessary.

But at this very moment in the film comes an event that I found quite jarring. When the lunch bag is placed before us, a small hand reaches, as if from behind us, to take it. The audience sees only the back of the hand, which we recognize as belonging to the boy—but I also immediately thought of operating it, as if it were a cursor in a videogame! Similarly, toward the end of the movie we are on a wonderfully realized street in contemporary Greenwich Village. It is a documentary shot—at street level, no spectacular helicopters, just life on that street corner at that moment. A couple in what would ordinarily be the background crosses the street. But there is no background. I am there. My attention is caught, and I want to follow that couple and see what *their* story is. Instead, the camera relentlessly drags me into a bar on the corner with the young boy. Again, I see a wonderfully detailed environment. Behind the bar are prints of some of the same stereopticon photos we have been seeing. I want to move closer, to lean into the shot and get a better view, but the camera stays with the dramatic action of the scene, namely, Tomas's conversation with the bartender. I am uncomfortable at these moments because the three-dimensional photography has put me in a virtual space and has thereby awakened my desire to move through it autonomously, to walk away from the camera and discover the world on my own.

The tension between watching a movie and being in a virtual place is even stronger in the more ambitious but less successful *Wings of Courage* (1995), a full-length IMAX feature that tells the story of the pioneer aviator Henri Guillaumet, who crashed his biplane in the Andes in 1930 and walked for six days and five nights through the snow to his rescue. Huxley's helicopter rules again in spectacular flight sequences that emphasize the fragility of the small planes against the vastness of the lonely mountains. But my immersion in these scenes was constantly disrupted by the director's shifting from interior to exterior shots and from one point of view to another. Such frequent cuts would be good practice for a conventional film (they would help the audience see the full picture), but they are out of place in a three-dimensional film, which can place me so concretely in space I become dizzy when shifting my point of view.

Again, it is the smaller places in the film that are the most arresting—a romantic period café, a cluttered office, Henri's girlfriend's cozy parlor. When the camera puts the audience at the same café table as the actors, the edge of the table is in the foreground and we can see to the left and right as well as across the table. When the waiter moves around the table, we see him from all angles. It is only when the camera angle switches that we are unpleasantly jarred from our trance of feeling that we are actually there.

Perhaps the most compelling environment in the film is the cave that Henri makes beside the wreckage of the plane. It is here that I experienced a surprising intimation of the dramatic potential of this medium. The hero Henri is describing, in voice-over, his plans for survival, carefully calculating the distance he must walk to safety and the time it will take to get there, as if he is writing in a pilot's logbook. His public voice is full of stoic resolve. But from the back of my headset comes a fearful whisper: "It can't be done. It simply can't be done." The filmmaker has taken me inside Henri's mind with startling effect. In some ways it is a Huxleyan moment. The audience is plugged into a sound machine, and it is goosing us. But in the context of the film, Henri's whisper of self-doubt is a moment of unmediated intimacy. It gave me chills not because of the gimmickry but because it brought me into unexpected closeness with this particular human being in his struggle for courage. At this one moment in an otherwise uninvolving story, I could sense the potential of this technology to take us seamlessly into a character's mind. The three-dimensional sound and images held out the possibility of a dramatic art form that can juxtapose the inner and the outer life as easily and gracefully as prose.

Riding the Movies

Huxley's fears are more fully realized in the sensation-oriented amusement park attractions that promise to let you "ride the movies." In this increasingly popular entertainment, the rider is placed on a hydraulically controlled movable platform or seat that tilts, twists, pitches, and shakes in synchronization with large moving images and environmental sound; an apparatus that seems very much like Huxley's pneumatic feely stalls. The concept of "riding" a movie fits the general strategy of entertainment industry conglomerates to create multiple "marketing windows" for the same imaginative product. If audiences loved to watch the DeLorean in *Back to the Future* or the motorcycle chases in *Robocop* or the magic carpet ride in *Aladdin*, they are primed to spend their money on rides based on these films. The first such attraction was the four-minute *Star Tours*, a ride developed in the early 1980s by two masters of cross-merchandising, Walt Disney Company and Lucasfilm. *Star Tours* was an immediate success.

The "movie ride" is engineered for strong visceral effects. It combines the surprises of the funhouse with the terrors of the roller coaster. According to Douglas Trumbull, who went from doing special effects in science fiction movies to making simulator rides, the aim is "to create an environmental total sensory experience that throws you right into the screen and you go into the movie."[8] As with three-dimensional films, the marketing emphasis is on the midway—bigger is better and biggest is best. So part of the attraction of Back to the Future, a ride that cost $16 million and uses three hundred speakers, twenty laser disc players, fifty miles of electrical wire, sixty video monitors, two 80-foot projection screens, and twenty computers, is that it is carefully engineered to provide the maximum thrill, to leave the rider breathless. "This ride can exert up to 1.8 Gs of force as it tilts and twists," says the Web page for the ride. "Compare the lowly airline jet, which rarely reaches 1.5 Gs!"

But the movie-rides are providing evidence that audiences are not satisfied by intense sensation alone. Once people do go "into" the movie, they want more than a roller-coaster ride; they want a story. Developers have lately been expanding the duration of the rides and are adding more characters and incidents to them to meet the rider's expectation of dramatic action. Most ambitiously, they are giving the rider more freedom to direct the ride and more opportunity to affect the unfolding story. The model is changing from one in which a rider is swept along in an exciting action to one in which a "guest" is paying a visit to an enticing place. For instance, on the Aladdin ride at Walt Disney World based on the animated feature film, you are seated on a magic carpet and allowed to move freely through the fantasy city of Agrabah. Because the developers had dynamically generated computer images rather than photographs, they were able to expand the world of film and to create an attraction that allows for multiple possible experiences. Guests are drawn into the town by the charm of its minarets, the mysteries of its back streets, and the presence of animated characters. They are given a role in the story, and their movements are motivated by the task of finding a hidden scarab. The Aladdin model suggests the possibility of a new kind of movie-ride, an adventure experience that is driven by the guests's curiosity and the beauty of the explorable world rather than by rushes of adrenaline.

Aladdin is an exception to the general trend, however, if only because of the high level of technical resources that Disney has poured into it, including special Silicon Graphics computers to generate the images in real time. For every one such attraction there will probably be hundreds of minirides based on limited movement, and much sketchier environments and focused on combat between customers within the virtual environment. Furthermore, the proliferation of even the high-end imaginative ride still raises the discomforting specter of a universe of entertainment products that advertise one another. See the movie! Ride the simulator! Play the game! The more successful such tactics prove, the more often movies will incorporate action sequences designed specifically for development as other "market windows." This may produce an entertainment paradise for fifteen-year-old boys, but it would mean an emotionally impoverished narrative form composed of many helicopter shots and far fewer moments of closeness with a particular human being.

Dramatic Storytelling in Electronic Games

While linear formats like novels, plays, and stories are becoming more multiform and participatory, the new electronic environments have been developing narrative formats of their own. The largest commercial success and the greatest creative effort in digital narrative have so far been in the area of computer games. Much of this effort has gone into the development of more detailed visual environments and faster response time, improvements allowing players to enjoy more varied finger-twitching challenges against more persuasively rendered opponents. The narrative content of these games is thin, and is often imported from other media or supplied by sketchy and stereotypical characters. This lack of story depth makes even wildly popular figures like the Mario brothers of the *Mortal Kombat* fighters impossible to translate into successful movie heroes.

In fact, in many maze-based games the story works against involvement in the game. One teenage fan of the X-Men, for instance, enjoyed the fighting moves of the characters in the *Clone Wars* game, which involves an invasion by the evil Phalanx, but found that the story line was inhibiting his ability to play. The game is structured so that the player is one of the X-men, who must save Earth from an invasion by the evil Phalanx forces. The X-men need the help of Magneto, a superhero who lives in a satellite stronghold. But in order to reach Magneto, the X-men must battle Magneto's soldiers in maze level after murderous maze level while receiving regular bulletins on the many countries that have fallen to the Phalanx. "Why should I want to kill these guys?" the player wanted to know. "We should all be working together." In order to make the conflict with the Phalanx the climax of the game, the developers had come up with a story of futile killing. As in many such games, the *Clone Wars* plot is contained in brief segments of text shown between the maze levels. The teenager wound up turning the story segments off altogether, as many players do with fighting games.

Electronic puzzle games rely less on violence than do twitch games. They also have a slower pace of engagement, since the player must figure out how to work the magic lever or where to search for the secret key. Although puzzle games can subordinate the story to the game play, just as the fighting games do, many puzzle games take advantage of this slower pace to offer a richer level of story satisfaction. In playing the early but still lovingly remembered text-based adventure game *Planetfall* (Infocom, 1983), you are a lowly deckhand on the spaceship *Feinstein*, which is soon destroyed by an explosion. Landing on a mysteriously deserted planet, you must survive long enough to figure out how to get away. In an abandoned laboratory, you find a deactivated robot, Floyd. Once you figure out how to turn Floyd on again, you are no longer alone. Wherever you go from then on within this baffling and dangerous world, Floyd is always there, chattering affectionately, begging for attention, playing with a rubber ball, and eagerly providing information and small services. After living through many adventures with Floyd, you reach the door of the radiation lab that contains a crucial piece of equipment. Inside the room are loud and dangerous mutants. As you stand outside the door listening to the murderous clamor, Floyd volunteers with characteristic childlike loyalty—"Floyd go get," he says—and rushes into the deadly chamber without giving you a chance to stop him. After accomplishing his mission, Floyd emerges "bleeding" oil and dies in your arms.

At this point the game changes from a challenging puzzle to an evocative theatrical experience. The escape from the planet continues, but without Floyd's company the player feels lonely and bereaved.

The memory of Floyd the Robot's noble self-sacrifice remains with players even years later as something directly experienced. "He sacrificed himself for me," is the way one twenty-year-old former player described it to me. Even those who speak of it less personally ("When you get to that room, he goes in to save you") convey a sense of wonder at the unexpected and touching quality of the gesture. The death of Floyd is a minor milestone on the road from puzzle gaming to an expressive narrative art. It demonstrates that the potential for compelling computer stories does not depend on high-tech animation or expensively produced video footage but on the shaping of such dramatic moments.

On the other hand, some game designers are making good use of film techniques in enhancing the dramatic power of their games. For instance, the CD-ROM game *Myst* (1993) achieves much of its immersive power through its sophisticated sound design. Each of the different areas of the game is characterized by distinctive ambient sounds, like the whistling of wind through the trees or the lapping of waves on the shore, that reinforce the reality of fantasy worlds, which are really just a succession of still images. Individual objects are also rendered more concrete by having them ping, thump, and whirr appropriately when manipulated. Wandering through a sinister fortress hideaway, I hear a musical motif that gets darker and more foreboding with each step and reaches an emotional peak when I uncover a severed head. The music track works as a game technique: it provides a clue that I am mouse-clicking along in the right direction, like the

hot and cold clues in a game of treasure hunt. But it is not gamelike
in tone. Instead, the solemnity of the music reinforces my feeling of
having come in immediate contact with a terrible act of depravity.
The music shapes my experience into a dramatic scene, turning the
act of discovery into a moment of dramatic revelation.

Games hold the potential for more powerful moments of revelation
than they currently make use of. Some years ago I was drawn into
playing a compelling arcade game while on vacation with my husband
and children. We had just entered the game room to give the kids a
treat, when I spotted a large-format TV screen in front of a laser gun
in the shape of a six-shooter. On the screen a cowboy was standing in
front of a low-cost version of the kind of TV Western set I spent much
of my childhood watching. "Howdy, partner," he said, and asked for
some help in running some bad guys out of town. I was immediately
hooked. It was clear to me that this was the game I'd been waiting for
all my life. I shot my way cheerily through the jail, saloon, livery stable,
and bank, knocking off the bad guys not quite as fast as the game
knocked off my supply of quarters. I was lost in a state of deep reverie.
Eventually my son and daughter ran out of quarters and came to find
me. As I turned toward them, I was conscious of being two very different people: the fervently pacifist mother who had taken them on
peace marches and forbidden all military toys and guns and the six-
shooting cowgirl who had grown up identifying with Annie Oakley
and Wyatt Earp. I would not claim that *Mad Dog McCree*, the game I
was playing, was a masterful piece of storytelling. But the moment of
self-confrontation it provoked, the moment in which I was suddenly
aware of an authentic but disquieting side of myself, seems to me to be
the mark of a new kind of dramatic experience.

Although economic and social forces may never move the established game industry far past the lucrative shoot-'em-ups and puzzle
mazes, there is no reason why more sophisticated developers could
not make stories that have more dramatic resonance and human im-
port to them, stories that, unlike Huxley's feelies, mean something,
just as Floyd's death is meaningful in the adventure game *Planetfall*,

the revelation of murder is meaningful in *Myst,* and the revelation of my own capacity for violence was meaningful to me in that arcade.

Story Webs

The accessibility of the World Wide Web has introduced a growing audience to hypertext fiction. Hypertext is a set of documents of any kind (images, text, charts, tables, video clips) connected to one another by links. Stories written in hypertext can be divided into scrolling "pages" (as they are on the World Wide Web) or screen-size "cards" (as they are in a Hypercard stack), but they are best thought of as segmented into generic chunks of information called "lexias" (or reading units).[9] Paper pages are bound into books in a single sequence; paper index cards must be arranged with no more than one card before and one after them even though they can be more easily searched in nonsequential order. But screen-based pages and cards become lexias: they occupy a virtual space in which they can be preceded by, followed by, and placed next to an infinite number of other lexias. Lexias are often connected to one another with "hyperlinks" (or "hot words"), that is, words that are displayed in color to alert the reader/viewer that they lead someplace else. For example, if I were writing this book as a hypertext, I would display the word *lexias* in the third sentence of this paragraph in color as a hot link instead of placing a superscript number next to it to indicate an endnote. Mouse-clicking on the word would bring up a new screen displaying the information on who invented the term and who applied it to electronic text, information that is now hidden at the back of the book. Another hyperlink might lead out of my book entirely and straight into a book by Roland Barthes or George Landow, or it could lead to a short bibliographical annotation that would pop up on the screen like a sticky note, appearing and disappearing at the will of the reader. A single lexia may contain many links, or it may contain no links at all, thereby gluing readers to the page or allowing them to move only forward or backward, as the pages of a book do. The existence of hypertext has given writers a new means of experimenting with segmentation, juxtaposition, and connectedness. Stories written in hypertext generally have more than one entry point, many internal branches, and no clear ending. Like the multiform life stories imagined by Borges and Lightman, hypertext narratives are intricate, many-threaded webs.

Hypertext formats are not new as intellectual structures. The Talmud, for instance, is a giant hypertext consisting of biblical text surrounded by commentaries by multiple rabbis. Literary works are hypertextual in their allusions to one another. In the twentieth century the allusiveness has grown so dense that a work like James Joyce's *Ulysses* is almost impossible to understand without accompanying pointers to other works, including a map of Dublin. *The Dictionary of the Khazars,* one of the multiform texts discussed earlier, is a print-based hypertext with entries that point to one another, making possible many coherent reading sequences. Although hypertext is not new as a way of thinking and organizing experience, it is only with the emergence of the computer that hypertext writing has been attempted on a large scale.

The hypertext formats of the 1990s support many kinds of narrative writing, from voyeuristic soap operas aimed at advertising revenues to postmodernist experimental fiction for university students. The first widely successful hypertext narrative is *The Spot,* a sexually titillating soap opera about a group of West Coast yuppies living in a beach house who post their diary entries regularly on the Web.[10] Readers can hop through the various diaries to compare different versions of the same event; can search through past events to catch up on the plot; and can even participate in the story by posting opinions, advice, or their own stories to a bulletin board in which the simulated characters participate along with fans. The characters in *The Spot* play to the prurient interests of the fans with a kind of self-mocking soft-core exhibitionism. For instance, in answer to one fan's challenge to prove that the diaries are being written in real time, a character posted a picture of herself, as directed, standing in a

bikini in front of the refrigerator and holding a strawberry. This cyberspace striptease, however appalling, is also indicative of the real
innovation behind this otherwise banal and poorly written soap. The
dramatic action is not in the canned story created by the writers
alone but in the spontaneously improvised exchanges between the
simulated characters and the participating fans. In defter hands such
audience engagement could provide imaginative, not just sexual, excitement.

The literary publisher Eastgate Systems distinguishes its products
from both pornographic "Web soaps" and games by calling them "serious hypertext." The pioneering work in this genre is Michael Joyce's
Afternoon (1987), written in the Storyspace hypertext system, which
he codesigned with Jay David Bolter and John Smith specifically for
the purpose of writing narrative as a set of linked text blocks. *Afternoon* contains 539 carefully crafted lexias and begins with one (although it does not necessarily come first) entitled "I Want to Say";
this lexia consists of a single compelling sentence: "I want to say I
may have seen my son die today." From here the reader is sent clicking through the cardlike lexia to find out more.

There is a lot to learn about the narrator, Peter, and about his ex-
wife, lovers, and friends, but most readers are not able to determine
whether his son is alive or dead or what Peter may have seen at the
site of a roadside accident. Instead, the reader circles through a complex web of lexia, each of which has several possible links to follow,
including a default "next" lexia, which appears in answer to a tap of
the return key. There is no overview of the work's structure, and the
"hot word" links do not offer much of a clue to the content to which
they lead. To complicate things further, Joyce has programmed some
of the links to force the reader to return to the same lexia again and
again in order to be permitted to go to new places in the story. This
continual circling through a confusing and contradictory space,
freighted with anxiety about the death of a child and irritation at
Peter's self-absorbed behavior, is reminiscent of a familiar *Star Trek*
plot—the one where the holodeck malfunctions; the characters act

out of role; and no matter what the crew members try, they cannot get out of the system.

But to the postmodernist writer, confusion is not a bug but a feature. In the jargon of the postmodern critics, Joyce is intentionally "problematizing" our expectations of storytelling, challenging us to construct our own text from the fragments he has provided. In the most praised effect of the story, he conceals a key section in a way that mirrors the protagonist's self-deceit. Only after repeated evasions can readers reach the lexia in which Peter will call his therapist and face his memory of his own culpability in the accident. For readers who enjoy the textured verbal labyrinth of *Afternoon,* there is a particular pleasure in coming to this section, although it does not have the finality of an ending or of an unambiguous solution to a mystery. Instead it deepens the range of possible interpretations of Peter's morning and afternoon.[11] The architectural playfulness of *Afternoon,* its construction as a series of discrete lexia linked by overlapping paths, and the poetic shaping of its individual lexia mark it as the first narrative to lay claim to the digital environment as a home for serious literature in new formats.

Much of the writing on the World Wide Web in 1996 is in standard short story format, perhaps with a few pictures or graphics added in; most writers have taken only limited advantage of the opportunity to write in hypertext structures. But the generation now in college grew up using encyclopedias on CD-ROMs and even making hypercard projects in the computer lab. In college, where they have an Internet connection that is faster than what they had at home, they use the World Wide Web as their primary source of reference material. They make their own hypertext self-portraits, in the form of personal "home pages," which they publish on the Web. Meanwhile, elementary and high schools are hooking up to the Internet in greater numbers every year. Unlike the first users of the medium, the next generation of writers will take the hypertext format for granted. As they come into greater expressiveness, they will bring the tangled structures of the current Web into more coherent order.

Computer Scientists as Storytellers

While dramatic and written narrative traditions have moved closer to the computer and computer-based entertainments have become more storylike, computer science itself is moving into domains that were previously the province of creative artists. Researchers in fields like virtual reality and artificial intelligence, who have traditionally looked to the military for technical challenges and funding, have recently turned from modeling battlefields and smart weapons to modeling new entertainment environments and new ways of creating fictional characters. These changes promise to greatly expand the representational power of the computer.

For instance, at Mitsubishi Electronics Research Laboratory researchers have created an appealing software environment that lets people at distant locations move through the same imaginary landscape. Diamond Park appears on large display screens as a grassy gathering place with bike trails, an outdoor restaurant, and inviting gazebos drawn in a vaguely turn-of-the-century style.[12] The bike trails are important because one of the first interfaces to this environment is a stationary bicycle equipped with a video display screen. You can move along the virtual trails by pedaling, just as you would move down a racecourse in an arcade driving game by stepping on the pedals of the car. But the difference here is that instead of racing forward, you can move in any direction (even off the paths), and the picture before you will change appropriately, reflecting your own position. You will also appear on the screen of other users, and they will appear on your screen as "avatar" figures (in this case, as animated drawings of people riding bicycles). Wearing a small microphone and headphones, you can talk to the other people as they come near; you can also pick up ambient sound, like music playing in the café. The bicycle interface acts like the vehicles in a movie-ride in that it makes the distances seen on the screen seem much more concrete by tying the visual movement to a kinetic environment. However, here the world is not built for adrenaline rushes but for socializing and exploration.

Sites like these (with or without bicycles) mark the future of the MUDs and chat rooms of the current Internet.

How present could we be in such environments? We could have our actual faces photographed in real time and mapped onto the avatars in the software. We could experience the virtual world not as a flat screen but as a virtual reality (VR) "pod" that surrounds us on six sides, like the holodeck. Although we would not have a holodeck chair to sit on, we could have something like a feely knob. We could wear clothing equipped with "tactors" that push back at us with the same pressure and texture as real objects. We could even hook the tactors up to distant objects, so that wearing a special glove we would "feel" the weight of an actual moon rock being lifted by a robot equipped with special sensors. Or we could hook up surgical instruments with tactors and attach them to a computer model of a patient, so that the images we would see would be reinforced by the appropriate feel of living tissue. Gamemakers are already adopting tactor technology to make more viscerally satisfying joysticks, and although the joysticks will not convey the sensation of a kiss, they will make for a more satisfying gun recoil or car collision.

Even without these force sensors, some VR installations of the 1990s are so visually present that interactors think they have touched things in the virtual world, including one another, even when they have not. One of the most intriguing such installations is the *Placeholder* world created by Brenda Laurel and Rachel Strickland for Interval Research Corporation in California.[13] Laurel, who holds the world's first Ph.D. in interactive narrative, has been designing games and user interfaces since the 1970s.[14] A critic of the conventional VR navigation system (in which users navigate by moving their hand or jiggling their head), Laurel designed an environment in which the system follows the changing full-body positions of people who move around in a natural way. Interactors wear VR helmets (which contain the three-dimensional visual display) and body sensors and must limit their movements to a "magic circle" marked out by rocks on the floor (an echo of the fairy ring, which is a traditional place of enchantment). Once inside the Placeholder world, they can enter the bodies of virtual animals and move as they move. For instance, if a woman in the crow's body spreads her arms, she sees her crow wings extend and her perspective changes as her crow body lifts off the ground. By swooping and banking appropriately she can take an exhilarating flight along a waterfall. Placeholder uses visual and sound motifs from the world of mythology to encourage collaborative imaginative play between pairs of interactors. It purposely avoids the commercial characters and weapon-driven competitive games of the movie-rides and arcade-style simulators. Placeholder suggests that reality technology can create a kind of stage set for adult improvisational play.

Perhaps the least encumbered holodeck experience available right now is in front of the twelve-foot computer screen set up by the ALIVE project of MIT's Media Lab as a "magic mirror" in which interactors see their own reflection placed beside the cartoon images of virtual characters designed in the lab.[15] In one scenario a little puppetlike child follows you around and tries to get your attention. In another a hamster scurries around, coming to you when you pick up some virtual food and hiding behind you when a foxlike predator is released. In a third a frisky dog named Silas will play fetch with you.[16] These attractive creatures live within the magic mirror as if it were a real three-dimensional space, an alternate reality echoing the rug area on which the interactor moves.

The Wonderland creatures on the other side of this looking glass are called "intelligent agents." They are computer-based characters with complex inner lives who can sense their environment, experience appetites and mood changes, weigh conflicting desires, and choose among different strategies to reach a goal. They are persuasively alive because their behaviors are complicated and spontaneous. They are quite life-size and they appear to be in real space with the interactor. Although they are still a very long way from Captain Janeway's romantic Lord Burleigh, such agents do have an independent existence of sorts and are significant steps on the road to believable holodeck characters.

When I play with Silas and his friends in front of the giant screen, they seem as alive as the animated figures in a movie—except that I am also in the movie. I have been prepared for this experience by watching so many movies that mix live and animated figures. However, it is much easier for me to suspend my disbelief in the existence of these creatures when someone else is interacting with me. The little puppetlike girl, for instance, came completely to life for me on the day when I was with a group showing the actress Lily Tomlin through the lab. Tomlin sat down on the carpet and patted the place beside her as the little figure shyly moved closer; the actress's gesture turned the interaction into a relationship, the beginning of a story of a developing intimacy. But seeing myself in the mirror, in my own ordinary clothes, which tell me I am in Cambridge rather than in Wonderland, I have a harder time sustaining the illusion.

Nevertheless, a floor-to-ceiling computer screen is an impressive way to display a virtual world. When the Media Lab setup is not in use for these advanced projects, graduate students play *Doom* by projecting its cavelike landscape on the screen and standing in front of it holding a plastic gun. The camera attached to the screen tracks the player's actions and sends messages to the game as if the player were holding a joystick. On the day I took a turn playing, the gun was not firing, but the fluid navigation through the enormous three-dimensional spaces was rapturous in itself.

In addition to creating vivid virtual worlds we can enter and fictional characters we can interact with, researchers are also developing complex computer models of plot. For instance, at Carnegie Mellon University, the Oz group, led by Joseph Bates, applies artificial intelligence techniques to storytelling.[17] One project of the group is based on an existing text-based computer game called *Deadline*.[18] Their goal is to customize the events of the murder mystery for each individual player so that the clues, red herrings, and revelations arrive at a satisfying pace, no matter what the player chooses to do. *Deadline* takes place in a mansion where there are suspects to be interviewed and physical evidence to discover. It is designed around a time scheme, so that if the detective does nothing to prevent it, a second murder takes place midway through the story. The Oz group analyzed all of the possible paths a player might take through the story and identified the ones that are the most satisfying. They then fed this information to a complex mathematical procedure called "adversary search," which is similar to the algorithms used in chess-playing systems, and which can calculate the optimal response to any action of the player in order to coax the player toward the most interesting narrative paths. A story system based on this design would eliminate the confused thrashing that accompanies much computer game playing by moving the interactor forward, not necessarily toward the solution to the puzzle but toward the most dramatically engaging encounters.

All of this research is still in the laboratory for now, but it is exciting to think about what it might add up to if all these technologies are combined. Imagine a visit to an entertainment venue of the late twenty-first century, the equivalent of a movie theater. The equivalent of a hit movie for the year 2097 might perhaps begin with a walk through a three-dimensional projected environment looking much like the theme-based restaurants and parks of our time or like the digital sets that are increasingly common in contemporary movies. We would be able to move the images around by moving our hands; for example, we might pick an illusory apple from a bowl of fruit or move an illusory chair. We would feel the weight and texture of these objects, although we could not eat the apple or sit on the chair. We would meet characters within this world who would sense our presence and converse with us; they would become as familiar to us as the characters in a beloved book or film. We would enter the story, and the plot would change according to our actions while still sustaining its power to surprise and delight us. What would such stories be like? How would we know what to do if we found ourselves inside one? Although we cannot predict how far the technology will take us, it is irresistible to speculate on such possibilities.

Even the near-term prospects are compelling. We are on the brink of a historic convergence as novelists, playwrights, and filmmakers move toward multiform stories and digital formats; computer scientists move toward the creation of fictional worlds; and the audience moves toward the virtual stage. How can we tell what is coming next? Judging from the current landscape, we can expect a continued loosening of the traditional boundaries between games and stories, between films and rides, between broadcast media (like television and radio) and archival media (like books or videotape), between narrative forms (like books) and dramatic forms (like theater or film), and even between the audience and the author. To understand the new genres and the narrative pleasures that will arise from this heady mixture, we must look beyond the formats imposed upon the computer by the older media it is so rapidly assimilating and identify those properties native to the machine itself.

Chapter 3

From Additive to Expressive Form

Beyond "Multimedia"

The birth of cinema has long been assigned to a single night: December 28, 1895. A group of Parisians, so the legend goes, were
gathered in a darkened basement room of the Grand Café on the
Boulevard des Capucines when suddenly the lifelike image of a
mighty locomotive began moving inexorably, astonishingly toward
them. There was a moment of paralyzed horror, and then the audience ran screaming from the room, as if in fear of being crushed by an
actual train. This no doubt exaggerated account is based on an actual
event, the first public showing of a group of short films that included
"Arrival of a Train at La Ciotat Station" by the Lumière brothers, who
(like Edison in America) had just invented a reliable form of motion
picture photography and projection. Film scholars have recently
questioned whether the novelty-seeking crowd really panicked at
all.[1] Perhaps it was only later storytellers who imagined that the first
projected film image, the novelty attraction of 1895, could have carried with it the tremendous emotional force of the many thrilling
films that followed after it. The legend of the Paris café is satisfying to
us now because it falsely conflates the arrival of the representational

technology with the arrival of the artistic medium, as if the manufacture of the camera alone gave us the movies.

As in the case of the printing press, the invention of the camera led to a period of incunabula, of "cradle films." In the first three decades of the twentieth century, filmmakers collectively invented the medium by inventing all the major elements of filmic storytelling, including the close-up, the chase scene, and the standard feature length. The key to this development was seizing on the unique physical properties of film: the way the camera could be moved; the way the lens could open, close, and change focus; the way the celluloid processed light; the way the strips of film could be cut up and reassembled. By aggressively exploring and exploiting these physical properties, filmmakers changed a mere recording technology into an expressive medium.

Narrative films were originally called photoplays and were at first thought of as a merely additive art form (photography plus theater) created by pointing a static camera at a stagelike set. Photoplays gave way to movies when filmmakers learned, for example, to create suspense by cutting between two separate actions (the child in the burning building and the firemen coming to the rescue); to create character and mood by visual means (the menacing villain backlit and seen from a low angle); to use a "montage" of discontinuous shots to establish a larger action (the impending massacre visible in a line of marching soldiers, an old man's frightened face, a baby carriage tottering on the brink of a stone stairway). After thirty years of energetic invention, films captured the world with such persuasive power and told such coherent and compelling stories that some critics passionately opposed the addition of sound and color as superfluous distractions.

Now, one hundred years after the arrival of the motion picture camera, we have the arrival of the modern computer, capable of hooking up to a global internet, of processing text, images, sound, and moving pictures, and of controlling a laptop display or a hundred-foot screen. Can we imagine the future of electronic narrative any more easily than Gutenberg's contemporaries could have imagined *War and Peace* or than the Parisian novelty seekers of 1895 could have imagined *High Noon?*

One of the lessons we can learn from the history of film is that additive formulations like "photo-play" or the contemporary catchall "multimedia" are a sign that the medium is in an early stage of development and is still depending on formats derived from earlier technologies instead of exploiting its own expressive power. Today the derivative mind-set is apparent in the conception of cyberspace as a place to view "pages" of print or "clips" of moving video and of CD-ROMs as offering "extended books." The equivalent of the filmed play of the early 1900s is the multimedia scrapbook (on CD-ROM or as a "site" on the World Wide Web), which takes advantage of the novelty of computer delivery without utilizing its intrinsic properties.

For example, one early version of a Web soap about a group of friends living in New York offers diary pages of text spiced with sexually suggestive photos. The wordiness of the journal keeps us constantly scrolling through the screens, impatient for something to happen in the narrated story or for something to do, like clicking on a link to get something new. There are, in fact, clickable buttons in the journal, but instead of offering new information they merely allow us to hear (after time delays for downloading the sound clip and for installing the necessary software to play the sound file if we do not already have it) actors speaking exactly the same dialog printed on the screen. The audio snippets are amusing novelties at best, and at worst they work like so many small apologies for the limits of the printed text. Just as the photographed plays of early filmmakers were less interesting than live theater, this early Web soap continually reminds us of how much less vivid it is than the romance novels and television dramas it draws upon.

A more digitally sophisticated Web soap would exploit the archiving functions of the computer by salting each day's new episode with allusions (in the form of hot word links) to exciting previous installments. Our clicking would then be motivated not by curiosity about the media objects (show me a video clip) but by curiosity about the plot (why does she say that about him?). The computer presentation would thereby allow pleasures that are unattainable in broadcast soaps. For example, we could follow a single appealing subplot while ignoring the companion plots that may drive us crazy, or we could come in at any time in the story and review important past events in all their dramatic richness. Instead of using audio redundantly to act out dialogue in a diary entry, a sophisticated Web soap might provide the audio as an integral part of the plotline—perhaps as the wiretap of a murder threat or a political negotiation or as a phone message that carries information of hidden romantic liaisons.

Some Web stories are already using such techniques, and no doubt all of them will in time. Their adoption is part of the inevitable process of moving away from the formats of older media and toward new conventions in order to satisfy the desires aroused by the digital environment. We are now engaged in thousands of such discoveries in all the subgenres of electronic narrative, the result of which will be the development of narrative pleasures intrinsic to cyberspace itself. Therefore, if we want to see beyond the current horizon of scrapbook multimedia, it is important first to identify the essential properties of digital environments, that is, the qualities comparable to the variability of the lens, the movability of the camera, and the editability of film, that will determine the distinctive power and form of a mature electronic narrative art.

ELIZA Comes to Life

Well before the advent of multimedia, there was a moment in the history of the computer that demonstrated its representational and narrative power with the same startling immediacy as the Lumières' train did for the motion picture camera. This is the famous but often misunderstood moment in which the first completely computer-based character was created.

In 1966, Joseph Weizenbaum, a professor of computer science at MIT, created, as an experiment in natural language processing, a computer program called ELIZA that carried on a conversation by replying to typed-in statements with printed words. Since this was before the widespread use of computer screens, the program used a form of teletype device connected to one of the first time-sharing computer networks. The resulting persona, Eliza, was that of a Rogerian therapist, the kind of clinician who echoes back the concerns of the patient without interpretation. She also displayed a narrowly Freudian interest in topics like sex and the family. In a research paper describing the program, Weizenbaum included this remarkably coherent sample conversation:

> *Patient:* Well, my boyfriend made me come here.
> *Eliza:* YOUR BOYFRIEND MADE YOU COME HERE.
> *Patient:* He says I'm depressed much of the time.
> *Eliza:* I AM SORRY TO HEAR YOU ARE DEPRESSED.
> *Patient:* It's true. I am unhappy.
> *Eliza:* DO YOU THINK COMING HERE WILL HELP YOU NOT TO BE UN-HAPPY?
> *Patient:* I need help, that much seems certain.
> *Eliza:* WHAT WOULD IT MEAN TO YOU IF YOU GOT SOME HELP?
> *Patient:* Perhaps I could learn to get along with my mother.
> *Eliza:* TELL ME MORE ABOUT YOUR FAMILY.[2]

Eliza proved astonishingly persuasive. Soon a creation legend started to circulate that was based on an incident involving a researcher at a Cambridge think tank who left an early version of ELIZA running on his computer, which a vice president of the company thought was typing out a communication being transmitted by his colleague from home. (Just as the Lumière train story is a staple of film history, the Eliza story in its several versions appears in many textbooks and computer science lectures, often told as if it happened to Weizenbaum himself.) The conversation takes place at nine on a Saturday morning:

> *Vice president:* Do you think that I can use the computer this morning?
>
> *Teletype:* WHY DO YOU ASK?
>
> *Vice president:* I might be able to make some additional telcomp (computer system) sales.
>
> *Teletype:* WHY AREN'T YOU SURE?
>
> *Vice president:* My prospective customers have never seen a demonstration of the system.
>
> *Teletype:* DOES THIS MEAN VERY MUCH TO YOU?
>
> *Vice president:* Of course it does.
>
> *Teletype:* YOU SEEM VERY POSITIVE ABOUT THAT.
>
> *Vice president:* This is ridiculous.
>
> *Teletype:* WHAT DO YOU MEAN THIS IS RIDICULOUS?
>
> *Vice president:* Is there any reason why I can't use the computer now?
>
> *Teletype:* TELL ME MORE ABOUT YOUR PROSPECTIVE CUSTOMERS.
>
> *Vice president:* Please dial me up on 491-1850

But because the vice president forgets to type in a final period, the machine does not answer him. Infuriated, he calls his colleague on the phone, resulting in this predictable exchange:

> *Vice president:* Why are you being so snotty to me?
>
> *Researcher:* What do you mean why am I being so snotty to you? (*Explosion of anger*)[3]

The story has become a legend because it discharges the anxiety aroused by the fear that Weizenbaum had gone too far, that he had created a being so much like an actual person that we would no longer be able to tell when we were talking to a computer and when to a human being. This is very much like the fear that people would mistake film images for the real world.

Eliza was not just persuasive as a live conversationalist; she was also remarkably successful in sustaining her role as a therapist. To Weizenbaum's dismay, a wide range of people, including his own secretary, would "demand to be permitted to converse with the system in private, and would, after conversing with it for a time, insist, in spite of [Weizenbaum's] explanations, that the machine really understood them."[4] Even sophisticated users "who knew very well that they were conversing with a machine soon forgot that fact, just as theatergoers, in the grip of suspended disbelief, soon forget that the action they are witnessing is not 'real' " (p. 189). Weizenbaum had set out to make a clever computer program and had unwittingly created a believable character. He was so disconcerted by his achievement that he wrote a book warning of the dangers of attributing human thought to machines.

Without any aid from graphics or video, Eliza's simple textual utterances were experienced as coming from a being who was present at that moment. What was the representational force that allowed the computer to bring her so compellingly to life?

The Four Essential Properties of Digital Environments

When we stop thinking of the computer as a multimedia telephone link, we can identify its four principal properties, which separately and collectively make it a powerful vehicle for literary creation. Digital environments are procedural, participatory, spatial, and encyclopedic. The first two properties make up most of what we mean by the vaguely used word *interactive*; the remaining two properties help to make digital creations seem as explorable and extensive as the actual world, making up much of what we mean when we say that cyberspace is *immersive*.

Digital Environments Are Procedural

Eliza was brought to life by the procedural power of the computer, by its defining ability to execute a series of rules. It is surprising how often we forget that the new digital medium is intrinsically procedural. Although we may talk of an information highway and of billboards in cyberspace, in fact the computer is not fundamentally a wire or a pathway but an *engine*. It was designed not to carry static information but to embody complex, contingent behaviors. To be a computer scientist is to think in terms of algorithms and heuristics, that is, to be constantly identifying the exact or general rules of behavior that describe any process, from running a payroll to flying an airplane.

Weizenbaum stands as the earliest, and still perhaps the premier, literary artist in the computer medium because he so successfully applied procedural thinking to the behavior of a psychotherapist in a clinical interview. It is the cleverness of Weizenbaum's rules that creates the illusion that Eliza understands what is said to her and that induces the user to continue the conversation. For example, if the user says, "Everybody laughs at me," the program can apply the rule that deals with *me* statements to echo the remark as, "You say that everybody laughs at you." This general rule models the neutrality of the Rogerian therapist, who reflects the patient's statements without judgment. Or, more cleverly, the program can select the specific rule for the word *everybody*, and reply, "Who in particular are you thinking of?"

Weizenbaum made sure the program would select the more specific reply by writing a rule that gives *everybody* a higher priority as a "key word" than *me*. Eliza's language processing includes no representation of syntax or meaning. Most of the time Eliza merely listens for key words such as *everybody* or *depressed* or *father*. Yet she is more believable than many programs with a more detailed representation of linguistic structure. Her credibility is the result of Weizenbaum's adroitness in formulating rules of discourse that are based on the ways in which a therapist would behave.

Eliza is not a neutral procedural model but a comic interpretation. Just as an actor might use makeup to emphasize a large nose or an owlish look, Weizenbaum used procedural notation to give parodic emphasis to some familiar absurdities of therapy itself. Like any classic type, Eliza is predictable in a way we instantly recognize and savor.

She is a descendent of the vaudeville doctor who bungles the exam, of the comic simpleton whose role is to misunderstand whatever is happening around her, and of the enjoyably repetitive flat characters in Dickens who repeat the same tag lines or behavior whenever we see them ("I never will desert Mr. Micawber," chimes the wife of the pompous screwup in *David Copperfield*).[5] The very rigidities of the computer are exploited for comic effect because they echo the rigidities of certain psychotherapeutic methodologies. The joke is on the therapists.

By mirroring the artificial constraints of therapy talk, Eliza makes them ridiculous. Consider, for instance, the following exchange:

> *Eliza:* Tell me more about your family.
> *Patient:* My mother is very motherly.
> *Eliza:* Who else in your family is very motherly?
> *Patient:* I am very motherly.
> *Eliza:* Is it because you are very motherly that you came to see me?
> *Patient:* Do you need a mother?
> *Eliza:* We are discussing you, not me.[6]

Eliza will reflect back any "I am X" statement with a formulation like "Is it because you are X that you came to see me?" Here, the patient responds to the suggestion by playfully challenging the shrink. But, Eliza invokes her rule for *you* questions, redirecting the conversation back to the patient like any well-trained therapist. Putting these two rules together creates a coherent dramatic dialogue with a punch line. Such dialogues are easy to generate and irresistible to attempt. They amount to a kind of collaboratively written comedy skit. Few people would now perceive Eliza as a real psychotherapist.[7] But as an improv partner, Eliza is still quite popular, available on most computer networks and sometimes sending a bill by e-mail after the session.

The lesson of ELIZA is that the computer can be a compelling medium for storytelling if we can write rules for it that are recognizable as an interpretation of the world. The challenge for the future is

how to make such rule writing as available to writers as musical notation is to composers.

Digital Environments Are Participatory

The energy with which people enter into dialog with Eliza is also evidence of a second core property of the computer: its participatory organization. Procedural environments are appealing to us not just
because they exhibit rule-generated behavior but because we can induce the behavior. They are responsive to our input. Just as the primary representational property of the movie camera and projector is
the photographic rendering of action over time, the primary representational property of the computer is the codified rendering of responsive behaviors. This is what is most often meant when we say
that computers are *interactive*. We mean they create an environment
that is both procedural and participatory.

Eliza's responsiveness is limited by her poor understanding of language, which makes her liable to nonsense utterances. Her direct successors are therefore mostly in research environments. It fell to
another group of MIT computer scientists to develop a fictional universe that structures participation more tightly, resulting in a more
sustained engagement.

A few years after the invention of ELIZA, researchers at the MIT
Laboratory for Computer Science brought forth a widely popular
computer-based story, the adventure game *Zork*, which is based on the
Dungeons and Dragons tabletop game.[8] In *Zork* the computer plays the
role of dungeon master by providing an invisible landscape that serves
as the game board and by reporting to players on the effects of their actions. Within *Zork*'s fantasy world, players move through dungeon
rooms by typing in navigational commands (north, south, east, west,
up, down), look for objects that can be manipulated (by typing appropriate commands, such as "read book," "take sword," "drink potion"),
solve riddles, and fight off evil trolls. The game (which, like ELIZA, is
still a popular feature of university networks) begins like this:

Welcome to Zork.
West of House.
You are in an open field west of a big white house with a boarded front door. There is a small mailbox here.

Interactor> Go north.
North of House.
You are facing the north side of a white house. There is no door here, and all the windows are barred.

Interactor> East.
Behind House.
You are behind the white house. In one corner of the house there is a small window which is slightly ajar.

Interactor> Open the window.
With great effort, you open the window far enough to allow entry.

Interactor> Go in.
Kitchen.
You are in the kitchen of the white house. A table seems to have been used recently for the preparation of food. A passage leads to the west, and a dark staircase can be seen leading upward. To the east is a small window which is open. On the table is an elongated brown sack, smelling of hot peppers.
A bottle is sitting on the table.
The glass bottle contains:
A quantity of water.

In making a fantasy world that responded to typed commands, the programmers were in part celebrating their pleasure in the increasingly responsive computing environments at their disposal. Before the 1970s most complex programming was done by writing a set of commands on a piece of paper; transferring them to keypunch cards; and taking the stack of cards to a mainframe computer (in an uncomfortably chilly room dedicated to keeping the machines from overheating), from which, much later, a cumbersome paper printout would emerge. Only one person could use a machine at a time. Whenever a program crashed (which was often), the output consisted of a "core dump"—a long series of 0's and 1's arranged in eight-digit units, showing what each bit and byte in the computer memory looked like at the moment the computer quit. Debugging a program in this environment was time-consuming and tedious.

In the mid-1960s research labs began developing the current computing environment of a display device and a keyboard (originally a telex machine) linked up to a time-sharing network that let programmers send input directly to a running program and receive a response. They were also making wide use of programming languages that were interpreted rather than compiled. All programming code written in higher-level languages (with commands like "If a = 1, then print file") must be translated into machine language instructions (with commands that look a lot like the raw 0's and 1's of the bits themselves) by either a compiler or interpreter program. Compiling your code before running it is like writing a book and then hiring someone to translate it for your readers. Using an interpreter is the equivalent of giving a speech with simultaneous translation. It provides more direct feedback from the machine and a more rapid cycle of trial and revision and retrial. The particular programming language in which both ELIZA and *Zork* were written, LISP (LIst Processing Language), was developed at MIT in the 1950s for artificial intelligence. Running LISP on a time-sharing system meant that its dynamic "interpreter" could immediately "return" an "evaluation" of any coded statement you typed into it, much as a calculator immediately returns the sum of two numbers. The result was a more conversational structure between the programmer and the program, a dialogue in which the programmer could test out one function at a time and immediately receive the bafflingly inappropriate or thrillingly correct responses. Both ELIZA and *Zork* reflected this newly animated partnership.

Whereas ELIZA captured the conversational nature of the programmer–machine relationship, *Zork* transmuted the intellectual challenge and frustrations of programming into a mock-heroic quest filled with enemy trolls, maddening dead ends, vexing riddles, and rewards for strenuous problem solving. ELIZA was focused on the cleverness of the machine-created world; *Zork* was focused on the experience of the participant, the adventurer through such a clever rule system. Zork was set up to provide the player with opportunities for making decisions and to dramatically enact the results of those decisions. If you do not take the lamp, you will not see what is in the cellar, and then you will definitely be eaten by the grue. But the lamp is not enough. If you do not take water with you, you will die of thirst. But if you drink the wrong water, you will be poisoned. If you do not take weapons, you will have nothing with which to fight the trolls. But if you take too many objects, you will not be able to carry the treasure when you find it. In order to succeed, you must orchestrate your actions carefully and learn from repeated trial and error. In the early versions there was no way to save a game in midplay, and therefore a mistake meant repeating the entire correct procedure from the beginning. In a way, the computer was programming the player.

Part of the pleasure of the participant in *Zork* is in testing the limits of what the program will respond to, and the creators prided themselves on anticipating even wildly inappropriate actions. For instance, if you type in "eat buoy" when a buoy floats by on your trip up a frozen river in the magic boat, then the game will announce that it has taken it instead and will add, "I don't think that the red buoy would agree with you." If you type in "kill troll with newspaper," it will reply, "Attacking a troll with a newspaper is foolhardy." The programmers generated such clever responses not by thinking of every possible action individually but by thinking in terms of general categories, such as weapons and foods. They made the programming function associated with the command word *eat* or *kill* check the player's typed command for an appropriate object; a category violation triggers one of these sarcastic templates, with the name of the inappropriate object filled in.

Because LISP programmers were among the first to practice what is now called object-oriented software design, they were well prepared to create a magical place like the world of *Zork*. That is, it came naturally to them to create virtual objects such as swords or bottles because they were using a programming language that made it particularly easy to define new objects and categories of objects, each with its own associated properties and procedures. The programmers also exploited a programming construct known as a "demon" to make some things happen automatically without the player's explicit action; for instance, in *Zork* a magic sword begins to glow if there is danger nearby, a stealthy thief comes and goes at his own will, and a fighter troll attacks the adventurer at unpredictable times. The programmers were also prepared by research on automatons to keep track of the state of the game, which allowed them to guess at the context of commands that would otherwise be ambiguous. For instance, if a player types "attack," the program looks around for a nearby villain and a weapon; if there are two weapons, it asks which one the player wants to use. These techniques, which were taken from simulation design and artificial intelligence work, allowed the *Zork* programming team to create a dynamic fictional universe.

By contrast, more conventional programmers of the 1970s were still thinking in terms of the branching trees, fixed subroutines, and uniform data structures that go back to the early understanding of the computer as a means of encoding information purely in the form of yes/no decisions. In fact, most interactive narrative written today still follows a simple branching structure, which limits the interactor's choices to a selection of alternatives from a fixed menu of some kind. The *Zork* dungeon rooms form a branching structure, but the magical objects within the dungeon each behave according to their own set of rules. And the interactor is given a repertoire of possible behaviors that encourage a feeling of inventive collaboration. The *Zork* programmers found a procedural technology for creating enchantment.

The company they formed, Infocom, is, though long out of business, still revered by players. Many fans attribute the imaginative superiority of Infocom games to the predominance of text over graphics, just as nostalgic radio fans prefer the sightless "theater of the imagination" to television. But though the writing in its games was skillful, it was not the true secret of Infocom's success. What made the games distinctive was the sophisticated computational thinking the programmers brought to shaping the range of possible interactions.

The lesson of *Zork* is that the first step in making an enticing narrative world is to script the interactor. The *Dungeons and Dragons* adventure format provided an appropriate repertoire of actions that players could be expected to know before they entered the program. The fantasy environment provided the interactor with a familiar role and made it possible for the programmers to anticipate the interactor's behaviors. By using these literary and gaming conventions to constrain the players' behaviors to a dramatically appropriate but limited set of commands, the designers could focus their inventive powers on making the virtual world as responsive as possible to every possible combination of these commands. But if the key to compelling storytelling in a participatory medium lies in scripting the interactor, the challenge for the future is to invent scripts that are formulaic enough to be easily grasped and responded to but flexible enough to capture a wider range of human behavior than treasure hunting and troll slaughter.

Digital Environments Are Spatial

The new digital environments are characterized by their power to represent navigable space. Linear media such as books and films can portray space, either by verbal description or image, but only digital environments can present space that we can move through. Again, we can look to the 1970s as the period that made this spatial property apparent. At Xerox PARC (Palo Alto Research Center) a group of visionaries created the first graphical user interface, the image of a desktop filled with file folders, which is currently the worldwide standard for computer file management. At Atari, inventors created the first graphics-based games, first *Pong* and then *PacMan*, which established the computer as a spatial game board and paved the way for today's videogames and CD-ROMs. At MIT the Architecture Machine Group, led by Nicholas Negroponte, created the first "surrogate travel system" by making a "movie map" of Aspen, Colorado, a photographic record of the town that could be "walked through" on the computer (each press of an arrow key or mouse click indicated a step forward) in simulation of an actual visit. All of these were important steps toward the current digital environment, as was the concurrent growth of the ARPANET (Advanced Research Projects Agency NETwork), the precursor of the Internet funded by the U.S. Department of Defense, which was mapping out a virtual space that was increasingly real to those who used it. We recognize the fruit of all of these developments in our conceptualization of the digital domain as "cyberspace," an environment with its own geography in which we experience a change of documents on our screen as a visit to a distant site on a worldwide web.

Although this spatial property has been widely exploited in graphical applications, it is in fact independent of the computer's ability to display maps, pictures, or even three-dimensional models. It is also independent of its communicative function in linking geographically distant places. The computer's spatial quality is created by the interactive process of navigation. We know that we are in a particular location because when we enter a keyboard or mouse command the (text or graphic) screen display changes appropriately. We can verify the relation of one virtual space to another by retracing our steps. The text-based dungeons of *Zork*, the sequenced stills of the enchanter's isle of *Myst*, the flat worlds of the multilevel maze games, the crow's waterfall visible in the Placeholder VR helmet, the continuous three-dimensional world of the new videogame dreamscapes— all are realized for the interactor by the process of navigation, which is unique to the digital environment.

Let us return to *Zork* to see how this navigational creation of space lends itself to dramatic engagement:

Interactor> West.
Living Room.
You are in the living room. There is a door to the east, a wooden door with strange gothic lettering to the west, which appears to be nailed shut, and a large oriental rug in the center of the room . . .

Interactor> Look under rug.
Underneath the rug is a closed trapdoor.

Interactor> Move the rug.
With a great effort, the rug is moved to one side of the room. With the rug moved, the dusty cover of a closed trapdoor appears.

Interactor> Open trapdoor.
The door reluctantly opens to reveal a rickety staircase descending into darkness.

Interactor> Go down.
It is pitch-black. You are likely to be eaten by a grue. Your sword is glowing with a faint blue glow.

Interactor> Turn on lamp.
The lamp is now on.
Cellar.
You are in a dark and damp cellar with a narrow passageway leading east, and a crawlway to the south. On the west is the bottom of a steep metal ramp which is unclimbable. The trapdoor crashes shut, and you hear someone barring it.

You, as player/interactor, have walked into a dungeon that someone is sealing shut behind you! The moment is startling and immediate, like the firing of a prop gun on the stage of a theater. You are not just reading about an event that occurred in the past; the event is happening *now*, and, unlike the action on the stage of a theater, it is happening to *you*. Once that trapdoor slams, the only navigational commands that work are the ones that lead further and deeper into the troll-filled lower world. The dungeon itself has an objective reality that is much more concrete than, for instance, the jail on the Monopoly board or a dungeon in a tabletop game of *Dungeons and Dragons*—or even a dungeon in a live-action role-playing game—because the words on the screen are as transparent as a book. That is, the player is not looking at a game board and game pieces or at a *Dungeons and Dragons* game master who is also in his or her algebra class or at a college classroom or campsite in the real world. The computer screen is displaying a story that is also a place. The slamming of a dungeon door behind you (whether the dungeon is described by words or images) is a moment of experiential drama that is only possible in a digital environment.

The dramatic power of navigation is also apparent outside the realm of the adventure game. For instance, Stephanie Tai, a student in my course on writing interactive fiction wrote a first-person poetic monologue about a sleepless night. Each screenful of text is a stanza and ends with a sentence fragment that connects syntactically with two or more stanzas, which are reached by clicking on arrows placed at the midpoint of the top, bottom, left, and right margins of the screen. Mouse-clicking through the mind of the insomniac is like a walk through a labyrinth. There are multiple end points to the maze, including one with just the single word *asleep* and another with the words *alone in this misery* in white letters on a black background. The poem is satisfying because the action of moving by arrows around a maze mimics the physical tossing and turning and the repetitive, dead-end thinking of a person unable to fall asleep. The movement through the cards makes a coherent pattern, but it is not one that could be modeled in physical space because the movement between links is not necessarily reversible. The navigational space of the computer allows us to express a sequence of thoughts as a kind of dance.

Stuart Moulthrop's ambitious hypertext novel *Victory Garden* (1992), whose title intentionally echoes the Borges story, is also in the shape of a labyrinth. Similar to a thick Victorian novel, it follows many characters with intersecting lives during the Gulf War. At the very center of Moulthrop's web is the death of Emily Runebird, an army reserve soldier who is killed in her barracks by an enemy missile. The attack itself is represented by a striking image of shattered text, as if the enemy shell itself had landed on the previous block of writing. We reach this image by following a continuous story thread, mouse-clicking through the screens automatically as if turning the pages of a book. The shattered screen stops us dead in our tracks. The effect of moving from the intact lexia to the shattered one is like an animation of the landing of the shell. The instant of time it takes to go from one screen to the other takes on a poignancy that reflects the abruptness of the soldier's death.

These very dramatic moments mark the beginning of a process of artistic discovery. The interactor's navigation of virtual space has been shaped into a dramatic enactment of the plot. We are immobilized in the dungeon, we spiral around with the insomniac, we collide into a lexia that shatters like a bomb site. These are the opening steps in an unfolding digital dance. The challenge for the future is to invent an increasingly graceful choreography of navigation to lure the interactor through ever more expressive narrative landscapes.

Digital Environments Are Encyclopedic

The fourth characteristic of digital environments, which holds promise for the creation of narrative, is more a difference of degree than of kind. Computers are the most capacious medium ever invented, promising infinite resources. Because of the efficiency of representing words and numbers in digital form, we can store and retrieve quantities of information far beyond what was possible before. We have extended human memory with digital media from a basic unit of portable dissemination of 100,000 words (an average book, which takes up about a megabyte of space in its fully formatted version) first to 65,000,000 words (a 650-megabyte CD-ROM, the equivalent of 650 books) and now to 530,000,000 words (a 5.3 gigabyte digital videodisc, equivalent to 5,300 books), and on upward.

Once we move to the global databases of the Internet, made accessible through a worldwide web of linked computers, the resources increase exponentially.

Just as important as this huge capacity of electronic media is the encyclopedic expectation they induce. Since every form of representation is migrating to electronic form and all the world's computers are potentially accessible to one another, we can now conceive of a single comprehensive global library of paintings, films, books, newspapers, television programs, and databases, a library that would be accessible from any point on the globe. It is as if the modern version of the great library of Alexandria, which contained all the knowledge of the ancient world, is about to rematerialize in the infinite expanses of cyberspace. Of course, the reality is much more chaotic and fragmented: networked information is often incomplete or misleading, search routines are often unbearably cumbersome and frustrating, and the information we desire often seems to be tantalizingly out of reach. But when we turn on our computer and start up our Web browser, all the world's resources seem to be accessible, retrievable, immediate. It is a realm in which we easily imagine ourselves to be omniscient.

The encyclopedic capacity of the computer and the encyclopedic expectation it arouses make it a compelling medium for narrative art. The capacity to represent enormous quantities of information in digital form translates into an artist's potential to offer a wealth of detail, to represent the world with both scope and particularity. Like the daylong recitations of the bardic tradition or the three-volume Victorian novel, the limitless expanse of gigabytes presents itself to the storyteller as a vast tabula rasa crying out to be filled with all the matter of life. It offers writers the opportunity to tell stories from multiple vantage points and to offer intersecting stories that form a dense and wide-spreading web.

One early indication of the suitability of epic-scale narrative to digital environments is the active electronic fan culture surrounding popular television drama series. As an adjunct to the serial broadcasting of these series, the Internet functions as a giant bulletin board on which long-term story arcs can be plotted and episodes from different seasons juxtaposed and compared. For instance, the Web site for the intricately plotted space drama *Babylon Five* contains images of the cast and plot summaries that document the many interwoven stories portrayed over multiple seasons, allowing a newcomer to understand the large cast of characters and the richly imagined array of alien races, each with its own culture and dramatic history. But it is not only science fiction programs that attract this interest. Even viewers of the mainstream television sitcom *Wings* use Web sites and Internet newsgroups to trace plot developments that extend over several years—like Joe and Helen's on-again, off-again courtship—and that may be confusingly jumbled in syndication; they also share digitized clips of favorite moments, such as the couple's comic wedding vows. The presence of such groups is influencing these shows, holding them to greater consistency over longer periods of time. In the past this kind of attention was limited to series with cult followings like *Star Trek* or *The X Files*. But as the Internet becomes a standard adjunct of broadcast television, all program writers and producers will be aware of a more sophisticated audience, one that can keep track of the story in greater detail and over longer periods of time. Since the early 1980s, when Steven Bochco introduced multiple story arcs with *Hill Street Blues*, television series have become more complex, involving larger casts and stories that take anywhere from one episode to several years to conclude. Some stories even remain open-ended after the series is over (especially if the writers were not expecting cancellation). In some ways, television dramas seem to be outgrowing broadcast delivery altogether. To join *Babylon Five* in its second or third season or *Murder One* in midseason is to immediately want to flip back or rewind to earlier episodes. The Internet serves that purpose, making a more capacious home for serial drama than the broadcast environment affords.

Making even fuller use of the computer's properties, by combining its spatial, participatory, and procedural elements with its

encyclopedic coverage, are the many on-line role-playing environments in the adventure games tradition. By the 1980s, Zork-like
games had grown to accommodate simultaneous multiple players,
turning them into Multi-User Dungeons or MUDs, which combine
the social pleasures of interplayer communication with the standard
command-driven adventures. In the MUDs of the 1990s players are
no longer limited to navigating a preexisting dungeon but can use a
simple programming language to build their own dungeon or adventure maze and link it up with those of other players by creating objects out of common building blocks. The MUD itself is a collective
creation—at once a game, a society, and a work of fiction—that is
often based on a particular encyclopedic fantasy domain, such as
Tolkien's Middle Earth or *Star Trek*'s twenty-fourth century. For instance, *TrekMuse*, founded in 1990 with over two thousand players,
had five hundred people enrolled in its virtual Star Fleet Academy in
1995, each of whom had made up his or her own character, based on
the existing *Star Trek* races. The digital narrative environment extends the fictional universe of the television shows and films in a way
that is consistent with the canonical version of the story but person-
alizes it for each of the players.

Some hypertext stories successfully use the encyclopedic extent of
the computer to develop multithreaded stories composed of many intersecting plots. In *Victory Garden*, for instance, we can follow a radical professor and his colleagues and graduate students through the
same time period as they intersect with one another in the class-
rooms, offices, and coffee bars, or we can follow them home to witness their tangled domestic lives; we can listen to the official
coverage of the Gulf War (with CNN transcripts) or read Emily
Runebird's letters. In *The Spot* and similar Web soaps, we can read
through the conflicting accounts of the same love affairs and decep-
tions in the journals of various friends. In on-line murder mysteries
like *Crime Story*,[9] we can delve through various document files, including crime scene photos, interview transcripts, and newspaper accounts. We can even leap out of the story altogether and find

ourselves in the "real" world, following a reference to the University of Mississippi right to its own Web site, or finding that the name of a witness seen in the company of the fleeing suspect belongs to a real-life software engineer whose Web page has nothing to do with the fictional crime. Not only does the weblike structure of cyberspace allow for endless expansion possibilities within the fictional world, but in the context of a worldwide web of information these intersecting stories can twine around and through the nonfictional documents of real life and make the borders of the fictional universe seem limitless.

However, the encyclopedic nature of the medium can also be a handicap. It encourages long-windedness and formlessness in storytellers, and it leaves readers/interactors wondering which of the several endpoints is *the* end and how they can know if they have seen everything there is to see. Most of what is delivered in hypertext format over the World Wide Web, both fiction and nonfiction, is merely linear writing with table-of-contents links in it. Even those documents designed explicitly for digital presentation, both fiction and nonfiction, often require too much superfluous clicking to reach a desirable destination or so much scrolling that readers forget where they are. The conventions of segmentation and navigation have not been established well enough for hypertext in general, let alone for narrative. The separation of the printed book into focused chapters was an important precondition of the modern novel; hypertext fiction is still awaiting the development of formal conventions of organization that will allow the reader/interactor to explore an encyclopedic medium without being overwhelmed.

The encyclopedic impulse and the dangers of the encyclopedic expectation are also apparent in simulation games. For instance, *Sim-City* (1987) presents the player with a schematic picture of a riverside city site, and places him or her in the role of mayor. The player is free to build the city however he or she would like, by adding to the model on the screen office buildings, factories, homes, a sewer system, electric power plants, a public transportation system, highways, schools, and so on. The software calculates the effects of each change by using models very like the ones used by social scientists and policymakers to study urban systems. Truly bad decisions in *SimCity* can bring critical newspaper articles, social unrest, and even electoral defeat. Well-built cities prosper through multiple decades. Because of the importance of the role in *SimCity*, the mayor is closer in power to God than to any real-life political leader, and the player's sense of omniscient awareness of consequences and omnipotent control of resources is part of the allure of such games.

Well-designed simulations like *SimCity* allow for multiple styles of play. One young programmer friend of mine spent hours building the most prosperous skyscrapered downtown possible. When I asked him about the game, he delighted in showing me the detail in which the city's underground service grid was specified. His wife, who is also a computer professional, took a different approach. Her favorite city was a sprawling environment with tree-lined family neighborhoods whose growing population gratified her tremendously and whose children she could easily imagine happily greeting each newly built playground. When they realized how much their efforts fell along gender lines, they laughed, but they pointed out that there was a more radical difference. For the husband, the program was a satisfyingly complex engineering problem, reinforcing his habitual sense of competence. For the wife, it was a narrative, in which the little parades and cheers of her contented townsfolk were the most memorable dramatic events. And, in fact, later versions of the game have been expanding this narrative quality by allowing the player to live inside a more detailed three-dimensional city rather than only manipulate it from on high.

Both the narrative possibilities and the godlike pleasures of simulation format are further developed in *Sid Meier's Civilization,* a game that puts the player in the role of leader of a civilization over the course of many centuries, while the computer plays the role of adversary civilizations that compete with the player for global resources and technical advancement. Like *SimCity, Civilization* allows multiple strategies of play and can accommodate the idealistic seeker of social harmony as well as the warrior player. The narrative interest of the game consists of creating multiple possible versions of an Earth-like history. For instance, it is possible to invent the railroad in B.C. times or to become an undefeated Napoleon. Winning the game is defined as either conquering all the other civilizations (in which case you are rewarded with pictures of the other leaders frowning at you) or sending twenty thousand people into space (in which case you see the spaceport).

Simulations like these take advantage of the authority bestowed by the computer environment to seem more encyclopedically inclusive than they really are. As its critics have pointed out, the political assumptions behind *SimCity* are hidden from the player.[10] This is less true in *Sid Meier's Civilization*, whose title alerts us to the fact that we are receiving a particular person's interpretation of human history rather than a scientific formula. The game also explicitly informs us that the behavior of each of the leaders is the result of three variables: their degree of aggression/friendliness, of expansionism/perfectionism, and of militarism/civilization. Since these are assumptions that players are aware of, they are free to accept or reject them as a reflection of the real world. Nevertheless, the basic competitive premise of the game is not emphasized as an interpretive choice. Why should global domination rather than, say, universal housing and education define the civilization that wins the game? Why not make an end to world hunger the winning condition? Why is the object of the game to compete with other leaders instead of to cooperate for the benefit of all the civilizations without jeopardizing any one country's security?

In an interactive medium the interpretive framework is embedded in the rules by which the system works and in the way in which participation is shaped. But the encyclopedic capacity of the computer can distract us from asking why things work the way they do and why we are being asked to play one role rather than another. As these systems take on more narrative content, the interpretive nature of these structures will be more and more important. We do not yet have much practice in identifying the underlying values of a multiform story. We will have to learn to notice the patterns displayed over multiple plays of a simulation in the same way that we now notice the worldview behind a single-plot story. Just as we now know how to think about what made Tolstoy propel Anna Karenina in front of that train or what made the producers of Murphy Brown offer her happiness as a single mother, we need to learn to pay attention to the range of possibilities offered us as interactors in the seemingly limitless worlds of digital narrative.

Digital Structures of Complexity

Like every human medium of communication, digital media have been developed to perform tasks that were too difficult to do without them. Hypertext and simulations, the two most promising formats for digital narrative, were both invented after World War II as a way of mastering the complexity of an expanding knowledge base. The mathematician Vannevar Bush put it this way in his landmark 1945 magazine article, "As We May Think": "The summation of human experience is being expanded at a prodigious rate, and the means we use for threading through the consequent maze to the momentarily important item is the same as that used in the days of square-rigged ships" (p. 102).

Bush's solution was "associational indexing" in a kind of magical desk based on microfilm files, a solution he called a "memex" and described as follows:

> The owner of the memex, let us say, is interested in the origin and properties of the bow and arrow. Specifically he is studying why the short Turkish bow was apparently superior to the English long bow in the skirmishes of the crusades. He has dozens of pertinent books and articles in his memex. First he runs through an encyclopedia, finds an interesting but sketchy article, leaves it projected. Next, in a history, he finds another pertinent item, and ties the two together. Thus he goes building a trail of many items. Occasionally he inserts a com-

ment of his own, either linking it into the main trail or joining it by a side trail to a particular item. . . . Thus he builds a trail of his interest through the maze of materials available to him.

And the trails do not fade. (P. 107)

This earliest vision of hypertext reflects the classic American quest—a charting of the wilderness, an imposition of order over chaos, and the mastery of vast resources for concrete, practical purposes. In Bush's view, the infinite web of human knowledge is a solvable maze, open to rational organization.

By contrast, Ted Nelson, who coined the term *hypertext* in the 1960s and called for the transformation of computers into "literary machines" to link together all of human writing, has been more in love with the unsolvable labyrinth. He sees associational organization as a model of his own creative and distractible consciousness, which he describes as a form of "hummingbird mind."[11] Nelson has spent most of his professional life in the effort to create the perfect hypertext system, which he has appropriately named *Xanadu*. He describes this pursuit as a quixotic quest, "a caper story—a beckoning dream at the far edge of possibility that has been too good to let go of, and just too far away to reach, for half my life."[12] Nelson's vision of hypertext is akin to William Faulkner's description of novel writing as a futile but noble effort to get the entire world into one sentence. Those like Nelson who take delight in the intricacies of hypertext, the twisting web rather than the clear-cut trail, are perhaps seeing it as an emblem of the inexhaustibility of the human mind: an endless proliferation of thought looping through vast humming networks whether of neurons or electrons.

The allure of computer simulations comes from a similar attempt to represent complexity. Three years after Bush's suggestion of the memex machine, Norbert Wiener founded the discipline of system dynamics with his book *Cybernetics*. Wiener observed that all systems, whether biological or engineered, have certain characteristics in common, such as the intertwining of multiple cause-and-effect relationships and the creation of feedback loops for self-regulation. Wiener called attention to parallels, for instance, between the way the body keeps a constant internal temperature by instituting changes (like sweating) and monitoring their effects (like feedback on skin temperature) and the way a home thermostat maintains a set temperature. Over the past fifty years, systems thinking has been applied to everything from family structure to frog ponds. It is now commonplace for us to think of the earth itself as a giant ecosystem, in both biological and political terms.

The computer has developed during this time into a versatile tool for modeling systems that reflect our ideas about how the world is organized. Early uses of computer simulations involved putting different values into a constant model and running the system through several "time steps" to see, for example, what would happen to crime statistics five, ten, and fifteen years down the line if police presence went up and cocaine prices went down. These systems were run in batch jobs, which spit out big chunks of numerical data. Other more responsive systems modeled a dynamically changing world open to real-time interaction, like the cockpit simulations used for training airplane pilots. In recent years, computer scientists have designed networked systems that are like a society full of autonomous individuals who talk and work with one another but have no single leader or controller.

In the late 1970s computer system design reached an intriguing milestone with a simple but elegantly conceived program that seemed to simulate life itself. The system is based on a checkerboard grid with markers that are white on one side and black on the other. The markers begin in a random arrangement and are then turned over according to a set of rules that makes decisions based on the colors of a marker's neighbors. Each round of turning causes more turning on the next round, eventually causing remarkable patterns to emerge and move across the board. The Game of Life system does not require a computer, but the patterns look particularly striking on the computer screen, which can run through multiple turns very quickly.[13]

Although no one would claim that such a system is alive in the same way as an animal or plant, it does capture one of the chief attributes of life—the creation of large patterns as a result of many smaller effects. Computer simulations like this are tools for thinking about the larger puzzles of our existence, such as how anything as soulless as a protein can give rise to something as complex as consciousness.

T. S. Eliot used the term *objective correlative* to describe the way in which clusters of events in literary works can capture emotional experience.[14] The computer allows us to create objective correlatives for thinking about the many systems we participate in, observe, and imagine. The rules for artificial life forms can be described as a kind of a game, but the knowledge about the world that the model offers us is not gamelike. It is a behavioral artifact that speaks to one of the most profoundly important aspects of our lives. The more we see life in terms of systems, the more we need a system-modeling medium to represent it—and the less we can dismiss such organized rule systems as mere games.

Current narrative applications overexploit the digressive possibilities of hypertext and the gamelike features of simulation, but that is not surprising in an incunabular medium. As digital narrative develops into maturity, the associational wildernesses will acquire more coherence and the combat games will give way to the portrayal of more complex processes. Participating viewers will assume clearer roles; they will learn how to become orienteers in the complex labyrinths and to see the interpretive shaping in simulated worlds. At the same time as these formal qualities improve, writers will be developing a better feel for which patterns of human experience can best be captured in digital media. In this way a new narrative art will come into its own expressive form.

The process by which this new art form will emerge is already under way and is itself interactive. Each time developers create new genres of digital stories or more immersive games, interactors try them out and grow frustrated or enchanted. Most often these incunabular products arouse expectations they cannot yet fulfill—for more encyclopedic coverage, for greater freedom of navigation, for more direct manipulation of the elements of the story. Every expressive medium has its own unique patterns of desire; its own way of giving pleasure, of creating beauty, of capturing what we feel to be true about life; its own aesthetic. One of the functions of early artifacts is to awaken the public to these new desires, to create the demand for an intensification of the particular pleasures the medium has to offer. Therefore, the next step in understanding what delights or dangers digital narrative will bring to us is to look more closely at its characteristic pleasures, to judge in what ways they are continuous with older narrative traditions and in what ways they offer access to new beauty and new truths about ourselves and the world we move through.

PART II

The Aesthetics of the Medium

ALAS, POOR YORICK! I KNEW HIM,
HORATIO: A FELLOW OF INFINITE JEST, OF
MOST EXCELLENT FANCY: HE HAT BORNE
ME ON HIS BACK A THOUSAND TIMES;
AND NOW, HOW ABHORRED IN MY
IMAGINATI0101000010101101011110101010
0101010010101100010101010101010100
0001010111010101010101001010010
01010101011110101010010111110101001
0011101001010001000101010010101
01000010111101010101000011110101
0101011101010101010101010111010100
010101010010111101010101000001011000
0101010101010111110101010101010101
0001010110101111010100101010100001
0101001010110100010101010101010000
001010111101010101010100101001010101
0101010111010101010010111101010011010
1110010101000100000101010010101010
0000101111010101010100000011101010
0101101010101010101010111101000000001
0101010101001110100100000101100010
1000101111010101011100100010101010101
1000101010010101010101000010111110101
01101010101010101010101111010101
0101011111010100000010101010100111100
0010000101100101000010111010101010
01010011111010101010000101010101010100
10101010100101010101111010101000010111111
0101001110011100101010000101000010
0101010101010000010111110101011100110
010101001010000101010101010101010000
0101111010101010000001111010101010101
01010101010101010111101000000101010
0100111001001000010110000101000
0111101010101010111101010101000101010
11010111101010101010101010010101010101
0101101000010101010101010000010101

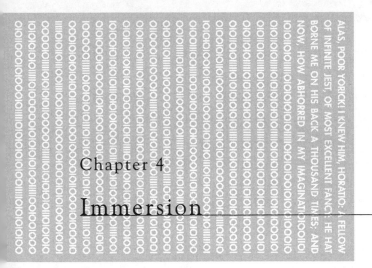

Chapter 4

Immersion

*In short, he so buried himself in his books that he spent
nights reading from twilight till daybreak and the days from
dawn till dark; and so from little sleep and much reading,
his brain dried up and he lost his wits. He filled his mind
with all that he read in them, with enchantments, quarrels,
battles, challenges, wounds, wooings, loves, torments, and
other impossible nonsense; and so deeply did he steep his
imagination in the belief that all the fanciful stuff he read
was true, that . . . [h]e decided . . . to turn knight errant
and travel through the world with horse and armour in
search of adventures.*

—*Don Quixote de la Mancha*

Don Quixote, living 150 years after the invention of the printing
press, exemplifies the dangerous power of books to create a
world that is "more real than reality." He still stands for the part of
each of us that longs to leap out of our everyday life into the pages of
a favorite book or, as the ride designers promise us today, to "go into

the screen" of a thrilling movie. A stirring narrative in any medium can be experienced as a virtual reality because our brains are programmed to tune into stories with an intensity that can obliterate the world around us. This siren power of narrative is what made Plato distrust the poets as a threat to the Republic. It is what made Cervantes' contemporaries fear the new fad of silent reading.[1] It is what made the advent of movies and television so frightening to the dystopian writers of the twentieth century. The same enchantment that sent Don Quixote tilting at windmills recently caused an Arkansas woman to show up for jury duty in the Whitewater case wearing a *Star Trek* uniform.[2]

The age-old desire to live out a fantasy aroused by a fictional world has been intensified by a participatory, immersive medium that promises to satisfy it more completely than has ever before been possible. With encyclopedic detail and navigable spaces, the computer can provide a specific location for places we long to visit. A few clicks on the World Wide Web and we are instantly in one of the feudal fiefdoms of the "current Middle Ages" set up by the Society for Creative Anachronism or in the sick bay of the starship *Voyager* being examined by the cranky doctor. Unlike Don Quixote's books, digital media take us to a place where we can act out our fantasies. With a telnet connection or a CD-ROM drive, we can kill our own dragons or fly our own starship; putting on a VR helmet or standing before a magic screen, we can do it all in 3-D. For the modern Don Quixote, the windmills have been preprogrammed to turn into knights.

The experience of being transported to an elaborately simulated place is pleasurable in itself, regardless of the fantasy content. We refer to this experience as immersion. *Immersion* is a metaphorical term derived from the physical experience of being submerged in water. We seek the same feeling from a psychologically immersive experience that we do from a plunge in the ocean or swimming pool: the sensation of being surrounded by a completely other reality, as different as water is from air, that takes over all of our attention, our whole perceptual apparatus. We enjoy the movement out of our familiar world, the feeling of alertness that comes from being in this new place, and the delight that comes from learning to move within it. Immersion can entail a mere flooding of the mind with sensation, the overflow of sensory stimulation experienced in the televisor parlor in Bradbury's *Fahrenheit 451*. Many people listen to music in this way, as a pleasurable drowning of the verbal parts of the brain. But in a participatory medium, immersion implies learning to swim, to do the things that the new environment makes possible. This chapter is about such digital swimming, about the enjoyment of immersion as a participatory activity.

Entering the Enchanted Place

The computer itself, even without any fantasy content, is an enchanted object. Sometimes it can act like an autonomous, animate being, sensing its environment and carrying out internally generated processes, yet it can also seem like an extension of our own consciousness, capturing our words through the keyboard and displaying them on the screen as fast as we can think them. As Sherry Turkle documents in her perceptive research on the psychology of cyberspace, working on the computer can give us uninhibited access to emotions, thoughts, and behaviors that are closed to us in real life.[3] MUDders and newsgroup members find it easy to project their deepest desires and fears onto people they have encountered only as words on a screen. People can fall in love very quickly over the Internet, and they also express their anger very easily (for example, by "flaming" one another in newsgroups). Some people put things on their home page (their site on the World Wide Web) that they have not told their closest friends. The enchantment of the computer creates for us a public space that also feels very private and intimate. In psychological terms, computers are liminal objects, located on the threshold between external reality and our own minds.[4]

Narrative is also a threshold experience. As we know from the work of child psychiatrist D. W. Winnicott, all sustained make-believe experiences, from children's play to Shakespearean theater, evoke the same magical feelings as a baby's first teddy bear because they are "transitional objects." The teddy bear provides comfort because the child projects upon it both his memories of the soothing mother and his sense of himself as a small being who can be cuddled and hugged. But though it embodies these strong subjective elements, the teddy bear is also a real object with a physical presence outside of anything the child imagines about it. To the baby it has a richly ambiguous psychological location, shimmering with emotion but definitely not a hallucination. A good story serves the same purpose for adults, giving us something safely outside ourselves (because it is made up by someone else) upon which we can project our feelings. Stories evoke our deepest fears and desires because they inhabit this magical borderland. The power of what Winnicott called "transitional" experiences comes from the fact that "the real thing is the thing that isn't there."[5] In order to sustain such powerful immersive trances, then, we have to do something inherently paradoxical: we have to keep the virtual world "real" by keeping it "not there." We have to keep it balanced squarely on the enchanted threshold without letting it collapse onto either side.

Because the liminal trance is so inherently fragile, all narrative art forms have developed conventions to sustain it. One of the most important ways they have done this has been to prohibit participation. Suzanne Langer, in her classic study of aesthetics, *Feeling and Form*, describes the "terrible shock" she received as a child watching a performance of James Barrie's *Peter Pan*:

> It was my first visit to the theater, and the illusion was absolute and overwhelming, like something supernatural. At the highest point of the action (Tinkerbell had drunk Peter's poisoned medicine to save him from doing so, and was dying) Peter turned to the spectators and asked them to attest their belief in fairies. Instantly the illusion was gone; there were hundreds of children sitting in rows, clapping and even calling, while [the actress], dressed up as Peter Pan, spoke to us

like a teacher coaching us in a play in which she herself was taking the title role. I did not understand, of course, what had happened; but an acute misery obliterated the rest of the scene, and was not entirely dispelled until the curtain rose on a new set. (Pp. 318–19)

Langer attributes her distress to the fact that art is dependent on establishing distance. To her mind, Barrie committed a theatrical sin by violating the fourth-wall convention that prohibits actors from acknowledging the spectators. The playwright's invitation to enter the circle of enchantment created by the stage is for Langer a shocking violation of the compact between playwright and audience. "To seek delusion, belief, and 'audience participation' in the theater is to deny that drama is art"(p. 319).

Whether or not it is destructive to art, audience participation is also very awkward. The literature of the twentieth century includes many concrete visions of the kind of boundary problems a truly participatory narrative would present. For instance, in Woody Allen's classic story "The Kugelmass Episode" a humanities professor at City College finds a magician with a kind of Don Quixote machine, a box that will allow him to jump into the pages of any novel he takes into it. Appropriately enough, Kugelmass chooses *Madame Bovary* and finds bliss with his fellow daydreamer by arriving just between her romances with Leon and Rodolfe. But students all over the country are confused: "Who is this character on page 100? A bald Jew is kissing Mme Bovary?" (p. 67). Kugelmass's problem is similar to the one I experienced standing in front of the magic mirror in MIT's Media Lab. When we enter the enchanted world as our actual selves, we risk draining it of its delicious otherness.

A simpler means to enchantment would be to bring to life a world that we wholly invent, a universal fantasy that is charmingly portrayed in Crockett Johnson's classic picture book, *Harold and the Purple Crayon*. Harold, a little boy drawn in black and white, carries a magenta crayon at arm's length as he walks across the pages of the book, drawing as he goes. Harold begins by improvising a sidewalk, an apple tree, and then a dragon to guard the apples. But the dragon scares him. His hand shakes and creates waves. He starts to drown in his own immersive world—until he thinks to draw a boat. In Johnson's fantasy, Harold's fluid imagination keeps getting him into and then out of such scrapes. External reality is represented by a black-and-white crescent moon that follows him no matter what he draws. At the end of his journey Harold becomes panicky when he cannot find his own room no matter how many buildings and windows he draws. Then he remembers that his window is always around the moon and realizes that he knows how to draw his way back into his own bed.

The digitally equipped Harold faces an intensification of Harold's perils. In the British space comedy *Red Dwarf*, a TV series, three unheroic space travelers—a fun-loving slob named Lister, a narcissistic humanoid evolved from a house cat, and the uptight moralistic Rimmer—receive a state-of-the-art "total immersion video" system based on mind reading. The game is called *Better Than Life*, and it is designed to immediately concretize the users' fantasies, like a sort of instantly programmable holodeck.[6] Lister and Cat joyfully imagine a motorcycle, a plush resort, and glorious meals, but the neurotic Rimmer finds himself unable to sustain pleasurable fantasies and involuntarily injects into their virtual paradise a tax collector, a deadly tarantula, and a torture scene with killer ants. End of game.

Ursula LeGuin pursues the same problem with more seriousness in her multiform novel *The Lathe of Heaven*. Here George Orr, an ordinary man, discovers that he has the magical power to remake reality literally according to his dreams. Despite his best intentions to save the world from disaster, George repeatedly awakens from dreams of peace and plenty to find that he has accomplished these ends by inflicting worse and worse catastrophes—from plague to alien invasion—upon his society. When he falls in love, George is tortured by the possibility that he will accidentally imagine a world in which his beloved is never born. The possibility of a magical domain in which our dreams can come true also arouses our most anxious nightmares.

The more present the enchanted world, the more we need to be reassured that it is only virtual and the more we need to see Harold's moon reminding us that there is a way back to the external world.

Participatory narrative, then, raises several related problems: How can we enter the fictional world without disrupting it? How can we be sure that imaginary actions will not have real results? How can we act on our fantasies without becoming paralyzed by anxiety? The answer to all of these questions lies in the discovery of the digital equivalent of the theater's fourth wall. We need to define the boundary conventions that will allow us to surrender to the enticements of the virtual environment. We cannot pick up the magic crayon until we have a clear fix on Harold's moon.

Finding the Border

Part of the early work in any medium is the exploration of the border between the representational world and the actual world. It is commonplace in the twentieth century to point to elaborate simulations of reality (electronic and otherwise) as a new and dangerous thing, a distancing of human beings from direct experience. But part of our dismay at televised events, wax museums, and immersive theme parks, at what Umberto Eco identified as the "hyperreal" quality of much of American life,[7] derives simply from the fact that we need time to get used to any increase in representational power. During this time one of our main activities, as creators and audience, involves testing for the boundaries of the liminal world.

At the beginning of the second part of *Don Quixote*, published ten years after the first, Cervantes has Don Quixote and Sancho Panza discuss the reception of the first part and quarrel with the representation of some of their adventures. Cervantes shows them meeting people who have read about them, thus mingling readers and fictional characters in the same illusory space. In the same way, characters on Web serials answer public fan mail and invite fans to post their own opinions and experiences to common bulletin boards. We get the same shiver from these posts today that Cervantes' readers experienced in his time. Just as we became accustomed to such devices in fiction, so too will we become used to them in cyberspace.

Similarly, when the form of the novel was beginning to coalesce in the eighteenth century, Laurence Sterne wrote a self-deconstructing memoir called *Tristram Shandy* in which the narrator inserts black pages, numbers chapters as if they had been rearranged, claims to have torn out certain pages, and sends us back to reread certain chapters. In short, he does everything he can to remind us of the physical form of the book we are reading. Sterne is exhilarated at his sheer power of representation, at the fact that he can transmit the voice of the imaginary Shandy into our minds using nothing but printed words. The brilliant animator Chuck Jones created at the height of his powers a similar virtuoso performance in *Duck Amuck*, which pits the pencil of a sadistic animator (revealed in the last frames to be Bugs Bunny) against an exasperated Daffy Duck. As Daffy tries to perform, the backdrop is redrawn from farm scene to castle to igloo; he himself is continually redressed, distorted, and even erased; the sound is divorced from the picture, so that guitars behave like machine guns; and the screen is allowed to go blank. After all the elements of cartooning have been deconstructed, Daffy is revealed to be in a filmstrip and two versions of himself confront one another from adjoining frames. The cartoon celebrates the persistence of the illusion. Just as Tristram Shandy survives a totally black page, Daffy Duck survives a totally white screen. Once the illusory space is created, it has such psychological presence that it can almost divorce itself from the means of representation.

Computer-based narratives are already showing the same tendency to emphasize the border, celebrate the enchantment, and test the durability of the illusion. In the experimental narrative installation *Archeology of a Mother Tongue*, produced for the Banff Center for the Arts in 1993 by Toni Dove and Michael Mackenzie, a key narrative transition takes the form of a system crash, which simulates a power failure in the virtual city represented by the surreal interface. Interactors must press a restart button on their screen to resume, and then find the city altered as if it had suffered a memory loss.[8] Even less artistically ambitious narratives offer similar effects. When my son puts down the game controller for a moment and pauses the action on the *Escape from Mars* maze game, the Tazmanian devil he had been controlling does not freeze in place. He glares out from the screen and begins to tap his foot and wave impatiently. This engaging comic gesture emphasizes the boundary between the puppet controlled by the player and the written character. It is almost as if the programmer within the system is waving at us, but doing so in a manner that deepens rather than disrupts the immersive world.

In the seventeenth and eighteenth centuries it was common to play with the borders of the illusion by presenting a novel as a collection of actual letters. Readers at the time were often confused (even two hundred years later I recently had a student who believed the fictional preface to *Les Liaisons Dangereuses* and accepted the exaggerated seduction stories as a true account). Early television shows like *Ozzie and Harriet* and *Burns and Allen* often fused the actor and the television character, suggesting to the audience that the virtual TV world was close to the actual lives of the stars. The premise was often accepted at face value by 1950s audiences, but few people watching the 1990s sitcom *Seinfeld* think that the comedian lives in New York rather than Los Angeles, where the show is produced, even though the character has the same name as the star. Web-based narrative is now playing the same kind of trick on us, by not giving us two sets of names to distinguish the actors from the characters they play and by linking fictional characters to sites in the real world.

Another way of exploring the border is to explicitly dramatize it. Winsor McCay, working at the very beginning of film animation in 1914, performed a vaudeville act in which he stood in a spotlight on stage and gave commands to a charming animated dinosaur, named Gertie, who appeared beside him on a giant movie screen. Gertie would have to be coaxed out slowly by him, but then she would perform tricks at his direction, snap at him when she got angry, and cry when he scolded her. At one point in the act McCay would take a prop cardboard apple, turn his back to the audience, and seem to throw it into the screen, where it appeared to land right in Gertie's mouth. At the dramatic climax of the act, McCay walked behind the screen and emerged as an animated drawing of himself. The animated McCay then stepped into Gertie's mouth so that she could lift him onto her back, where he took his bows while Gertie gracefully carried them both offscreen.[9]

The difference for the audience between the boundary experiments of earlier media and the ones that artists are now undertaking in the digital world is that this time *we* have also been invited into the mouth of the dinosaur.

Structuring Participation as a Visit

How will we know what to do when we jump into the screen? How will we avoid ripping apart the fabric of the illusion? Participation in an immersive environment has to be carefully structured and constrained. Ideally, the range of allowable behaviors should seem dramatically appropriate to the fictional world, just as ELIZA structured conversation in the format of a psychiatric interview and *Zork* constrained responses to the adventure game. For purposes of experiencing multisensory immersion, one of the simplest ways to structure participation is to adopt the format of a visit. The visit metaphor is particularly appropriate for establishing a border between the virtual world and ordinary life because a visit involves explicit limits on both time and space.

Amusement park fun house rides are a familiar model for an immersive visit that is also a narrative. The fun house has an entrance and an exit that mark the beginning and end of the story. As the visitor progresses on a moving platform, the dramatic tension builds from small surprises and hints of danger; then there are thrills and a mounting sense of threat or terror, which culminates in a big finish such as a free fall or an attacking beast. Like a movie set or theatrical stage, the fun house ride is calculated to look as if it had a fuller existence, even though the illusion is meant to be seen only from a particular angle and in carefully timed momentary glimpses. A fun house is a movie made into a machine that you travel through.

Most amusement rides still assume that the visitor can do nothing more than sit and scream. But that does not mean that they are easier to make than movies. For instance, most of the dinosaurs in the movie *Jurassic Park* were part of a virtual set; computer models were drawn, rather than built, and then transferred directly to the film. Those that were built were only partial dinosaurs, meant to be photographed from one angle at a time. By contrast, the spectacular Jurassic Park attraction at the Universal Studios theme park has to be much more explicit. Its models are giant dinosaur-size robots that move realistically on special hydraulic cylinders designed to produce a smooth motion. They are made to be viewed from multiple angles and have special realistically textured skin that clings and flaps from the robot's metal frame. The amusement ride occupies five acres and accommodates three thousand visitors per hour on its twenty-five-person boats. The various events of the ride—the surprising appearances of the various dinosaurs, the flashing of warning lights, the glimpses of an overturned jeep, the attack of the dinosaurs, the destruction of the breeding lab—unfold as the boat passes the corresponding trigger point. Unlike the video-based *Back to the Future* or the graphics-based ride *Aladdin* rides (described in chapter 2), the Jurassic Park ride seems like a visit to a real place. The visitor even gets wet during the eighty-four-foot plunge that gives the ride its big finish. But Jurassic Park is not a place, any more than a theatrical stage is, since a visitor cannot step off the boat without destroying the experience. Jurassic Park is essentially a giant computer-driven machine for telling an immersive story, and the boat is the fourth wall, an enchanted threshold object that carries you into the immersive world—and then out again. Like Harold's moon, the Jurassic Park boat is both part of the illusory world and also a reminder of the boundaries. Sitting within it, you are free to give way to terror without worrying about being able to find your way back.

Screen-based electronic environments can also provide the structure of an immersive visit. Here the screen itself is a reassuring fourth wall, and the controller (mouse or joystick or dataglove) is the threshold object that takes you in and leads you out of the experience. When the controller is very closely tied to an object in the fictional world, such as a screen cursor that turns into a hand, the participant's actual movements become movements through the virtual space. This correspondence, when actual movement through real space brings corresponding movement in the fantasy world, is an important part of the fascination of simple joystick-controlled videogames. Moreover, an electronic game that involves a maze and combatants is also very much like a fun house visit in that opponents keep popping out at you and obstacles keep appearing in your path in a randomized and therefore surprising fashion. This constant activity means that even if you move through the space without fighting, the world is still dramatically present; this is not a passive game board but a live-action stage.

By contrast, one of the limitations of the graphically immersive world of *Myst* is that it is dramatically static. Nothing happens of its own accord as the player wanders around in search of puzzles to solve. *Myst* sends us on a treasure hunt in a weirdly depopulated environment, a quest that is only partially motivated by the story. The lack of dynamic events reflects the simplicity of the underlying programming. *Myst* offers the interactor an elegant and seamless interface in which most of the activity of the game is moving forward through a space by mouse-clicking in the direction you wish to go. There are no enemies to encounter or people to bargain with. Few of the puzzles require any carrying of objects from one location to another. *Myst* is an unusually nonacquisitive and nonviolent game compared to most puzzle quests. The solution to the puzzles often depends on subtle aural cues, increasing the player's attentiveness to the meticulous sound design. In short, there is almost nothing to distract you in *Myst* from the densely textured visual and aural environment, but this intense immersion in visiting the place comes at the cost of a diminished immersion in an unfolding story.

The visitor role is also exploited in the CD-ROM version of the starship *Enterprise*, a "technical manual" that promises to use "a subset of holodeck technology" to present the starship and that includes a voice-over tour from Commander Riker. The visuals are produced from video of the key sets from the TV series *Star Trek: The Next Generation* and processed with a virtual reality tool (QuickTime VR) that lets you rotate your onscreen position 360 degrees and step forward and backward within continuous space, a tremendous improvement over the discontinuities of still-frame representations like those in *Myst*. The movement is so fluid, the visuals have such authority, and the representation is so complete that our visit to the *Enterprise* has a magical quality; it is as if we are aboard the real starship, the canonical location of the fictional world of which the television and movie representation are just copies. But after we check out all the key places—the captain's ready room, the bridge, the lounge area on 10-Forward, the quarters of all the crew members— the visit to the *Enterprise* loses its immersive hold because nothing is happening there. In a digital environment we do not want to use a spaceship as a databank. The more we feel that we are actually there, the more we want to fly off on it and have adventures.

In environments based on the amusement park model, the story and the visit can be tightly meshed. Objects can perform for us as we pass in front of them, their performance triggered by our presence. But if the interactor is not allowed to step off the moving platform, the visit will have to be short and full of intense stimulation to hold our attention and keep us from wanting to go off to explore the space. A more exploratory visit, on the other hand, can feel very lonely without other characters to engage with or a drama that unfolds in real time. Because we experience ourselves as present in these immersive worlds, as if we are on the stage rather than in the audience, we want to do more than merely travel through them.

The Active Creation of Belief

The pleasurable surrender of the mind to an imaginative world is often described, in Coleridge's phrase, as "the willing suspension of disbelief." But this is too passive a formulation even for traditional media. When we enter a fictional world, we do not merely "suspend" a critical faculty; we also exercise a creative faculty. We do not suspend disbelief so much as we actively *create belief.* Because of our desire to experience immersion, we focus our attention on the enveloping world and we use our intelligence to reinforce rather than to question the reality of the experience.

As the literary theorists known as the "reader response" school have long argued, the act of reading is far from passive: we construct alternate narratives as we go along, we cast actors or people we know into the roles of the characters, we perform the voices of the characters in our heads, we adjust the emphasis of the story to suit our interests, and we assemble the story into the cognitive schemata that make up our own systems of knowledge and belief. Similarly when we watch a movie, we take the separate spaces of the various sets and merge them into a continuous space that exists only in our minds. We take fragmentary scenes and mentally supply the missing actions; if someone is seen with a grocery bag and then working over a stove, we understand the meal is effortful. If someone is wearing an Ivy League sweatshirt, we might assume they are intelligent and earnest or maybe spoiled and preppy. We bring our own cognitive, cultural, and psychological templates to every story as we assess the characters and anticipate the way the story is likely to go.[10]

In a complex narrative world we can reinforce our belief by writing scholarly analyses or fanzine articles that analyze the underlying assumptions of the world, whether they concern Irish history or matter replicators. Encyclopedic writers like James Joyce, Faulkner, Tolkien, or Gene Roddenberry evoke this kind of response by the encyclopedic detail and intricacy with which they present their fictional creations. Such immersive stories invite our participation by offering us many things to keep track of and by rewarding our attention with a consistency of imagination.

In digital environments we have new opportunities to practice this active creation of belief. For instance, in an interactive video program set in Paris that my research group designed in the 1980s for language learners, we included a working telephone, represented by a photograph of a phone whose keypad could be clicked on. Students found the phone in an apartment they were free to explore by stepping through a photographed space. Near the phone were the numbers of people they had been motivated to telephone by the plot of the story (and whose answering machines they reached when they called). If they punched in a number outside the game, they heard the authentic out-of-service message used in Paris. The story was mostly told in well-directed video segments, which the students also found enjoyable, but the telephone was one of the most popular features of the story. This was because it behaved as a functional virtual object and because it became part of the accomplishment of a specific goal. In short, it became real through use.[11]

In the CD-ROM game *Star Trek: The Final Unity*, the player has to figure out how to free a woman scientist trapped under a pipe after an attack on a power plant. The pipe is too heavy to lift and it cannot be vaporized with the crew's phaser guns. The solution is to use a tricorder to record the coordinates of the pipe's location and then go down to the transporter room on the first floor to enter the coordinates into the transporter to "lock onto" the pipe and beam it off of her. If this is done right, the pipe appears in the transporter room, materializing to the accompaniment of the familiar tinkling transporter sounds. Operating the tricorder and the transporter in this way— which really only means clicking the mouse here and there on some unspectacular screen graphics—makes the world of the game seem much more present than does the same world on *Starship Enterprise,* the more visually impressive CD-ROM. It is the experience of using the objects and seeing them work as they are supposed to in our hands that creates the feeling of being a part of the *Star Trek* world.

The great advantage of participatory environments in creating immersion is their capacity to elicit behavior that endows the imaginary objects with life. The same phenomenon occurs when a child rocks a teddy bear or says "Bang!" when pointing a toy gun. Our successful engagement with these enticing objects makes for a little feedback loop that urges us on to more engagement, which leads to more belief. As the digital art medium matures, writers will become more and more adept at inventing such belief-creating virtual objects and at situating them within specific dramatic moments that heighten our sense of immersed participation by giving us something very satisfying to do.

Structuring Participation with a Mask

Cyberspace gains much of its immersive power from spectacular effects—arresting visuals like the fast-moving, pulsating explosions of the videogame, the flashing billboards of the World Wide Web, and the hallucinatory apparitions of virtual reality landscapes. This visual pageantry links computer culture to ancient forms of entertainment. Spectacle has traditionally marked the descent into a gathering of ordinary mortals of a godlike being—Dionysus, a Hopi kachina, the pope during a papal procession, a royal bride and groom, or Santa Claus rolling down Broadway to Macy's department store every Thanksgiving Day. Spectacle is used to create exultation, to move us to another order of perception, and to fix us in the moment.[12]

Historically, spectacle tends to moves toward participatory narrative in order to retain our attention, to lengthen the immersive experience. For instance, in the Middle Ages, the rituals of the church were extended into a folk dramatic form. Mystery, or miracle, plays were performed on wagons that rotated around the town; each episode was staged by an appropriate guild, with shipbuilders doing the story of Noah and cooks using their pots and pans to simulate the clatter of the Harrowing of Hell. The tradition survives today in parade floats and in the Nativity pageants still popular at Christmas. Renaissance masques, a secularized form of pageantry, were often performed by aristocratic guests at celebrations that ended with an unmasking and general dance. In the twentieth century, Halloween is widely celebrated as a giant participatory costume pageant. True to the ancient origins of the holiday, there are processions of costumed figures and a large component of neighborly participation.

In all of these traditions, participation in the spectacular event begins with ordinary people, rather than professional entertainers, donning a costume or mask. The mask sets off the participants from the nonparticipants and reinforces the special nature of the shared reality. It creates the boundary of the immersive reality and signals that we are role-playing rather than acting as ourselves. The mask is a threshold marker, like Harold's moon or the Jurassic Park boat. It gives us our entry into the artificial world and also keeps some part of ourselves outside of it.

In digital environments we can put on a mask by acting through an avatar. An avatar is a graphical figure like a character in a videogame. In many Internet games and chat rooms, participants select an avatar in order to enter the common space. Even when avatars are crudely drawn or offer a very limited choice of personalization, they can still provide alternate identities that can be energetically employed. For instance, the inclusion of graphic avatars in the networked action game called *Quake* led players to organize themselves into clans. Each clan dresses its avatars in the same colors, and its members fight together against other clans. *Quake* players have created an array of clan web pages, which look like what the Crips and Bloods might create if they traded their semiautomatics for laptop computers.

Virtual reality technology can offer a new kind of costuming and pageantry. Brenda Laurel and Rachel Strickland have devised "smart costumes" for the virtual playground called *Placeholder* (described in chapter 2). In fact, the participants are doubly costumed, since they are wearing actual helmets and body sensors that allow them to enter the virtual animal bodies that make up the smart costumes within the imaginary world. The virtual costumes are "smart" in that the interactor's vision, voice, and movement change appropriately as he or she changes, for example, from a swimming fish to a hissing, slithering snake. Since the system is designed for two players to inhabit the imaginary worlds together, they can enjoy the pleasures of a masquerade by showing off their costumes to one another and observing each other's displays. Participants are so present in the space that they sometimes think they have touched one another, even though they are actually physically isolated and unconnected by tactile sensors. Since *Placeholder* is based on a childhood model of play in which the interactors make up their own stories, the smart costumes are a kind of dress-up box, a set of enchanted story materials that provide a stimulus for improvisation.

There is a similar pleasure in embodiment in the Oz group's screen-based *Woggles* creatures at Carnegie Mellon University.[13] Here the user is invited to operate a cartoon figure with large eyes and an oval, stretchy body that can leap and slide and bow through a simple two-dimensional graphics world in the company of other creatures who behave autonomously. Since Woggles are programmed to play together and imitate one another, once you learn how to make your creature slide, another creature may slide after you. This world is engaging for people who do not like to operate the characters in fighting games; here the object is not to master a set of joystick twitches in order to destroy an opponent but to participate in a social world by taking on an intriguingly flexible body whose movements are also a means of communication. Entering a Woggle body is like becoming a citizen of Woggleland. It is as if you could put on a beret and start to shrug, gesture, and even speak like a Frenchman.

Smart costumes and social avatars are encouraging steps in the direction of a more expressive and less gun-crazy medium.

Structuring Collective Participation with Roles

The presence of other participants poses special challenges to immersion. For Suzanne Langer, the other children in the audience watching *Peter Pan* disrupted her immersion in a shocking way. But this is not a necessary effect. Like many baby boomers, I first experienced *Peter Pan* not in a theater but on television. I can vividly remember the thrill I felt, sitting on the floor in my living room close to the screen, when Mary Martin's Peter looked into the camera and asked us all to clap for Tink. I felt part of a vast effort that was truly healing her as I clapped away. But I also remember my self-consciousness in subsequent showings when my parents expressed amusement at my reaction. The problem for me was not with Peter Pan turning to the camera but with my awareness of unbelievers in the rest of the audience.

Clapping for Tinkerbell disturbed Langer in part because it is too explicit an enactment of the audience's role in sustaining the theatrical illusion. By gathering together in a theater, maintaining silence, and applauding in ritual ways, the audience creates the magic spotlight in which the actors move. But when Peter makes our applause a direct expression of belief in the imaginary, we are then reminded of the fact that Tinkerbell is only a trick of lighting on a stage. Perhaps the ideal way to clap for Tink is to do so alone in a room with a television set, aware of all the other people watching and clapping but not actually hearing them. This is the experience of the MUDs.

The power of a MUD is that the computer filters out the distraction of the actual appearance of the other players who are present. What is visible instead is their assumed identity, the role that everyone must choose in order to log on to the MUD. When you join a MUD, you assign yourself a sex and a physical description; if it is a very structured game, you acquire a set of attributes and skills represented by numerical values (e.g., magical powers = 10, strength = 8). As Sherry Turkle has pointed out, people do not so much play in MUDs as move into them.[14] They can sustain a role over a long period of time, accumulating experience points in a structured game by killing trolls or finding treasure or by learning to pilot a starship. Or they can just accumulate social experience in role-playing a particular kind of character—a scheming necromancer or a hyperrational Vulcan. In very story-specific MUDs, crucial roles such as the role of the wizard Gandolf in a Tolkien MUD may only be available by audition, but most MUDs allow the players to invent their own characters within the conventions of the controlling fictional genre. The role is therefore a combination of personal fantasy and collectively recognized conventions.

One key to functioning in a MUD is the ability to flip back and forth between player and character, to remove the mask in order to adjust the environment and then to put it back on again. For instance, if a player becomes frustrated with someone who is being too intransigent in a negotiation, he or she might send the following double message:

> IN [in character]: Please consider withdrawing your ultimatum.
> OOC [out of character]: Just because you're a Klingon doesn't
> mean you have to act like a jerk.

Sharing an unscripted fantasy environment with other people entails a constant negotiation of the story line and also of the boundary between the consensual hallucination and the actual world. When things are going well, the players can provide one another with a collective creation of belief that is like the shared make-believe of childhood. But when it is going badly, the player is stuck with a sputtering story line from lack of consensus or is left stranded with no one logged on to play with.

In the view of some players, live-action role-playing games (LARP) offer more coherent stories than MUDs.[15] Because the players are visible to one another and clearly not in a spaceship or a medieval castle (but, probably, in the basement of a university or the cabin of a summer camp), live-action games rely on explicit mechanisms of participation to sustain the illusion of a fictional world. One of the most powerful strategies, used by the role-playing group at MIT for instance, is the development of specific character profiles by the game masters to guide the individual players without rigidly prescribing their actions. The character profiles, provided to the players in advance of the game, are a combination of background story and game goals. In the hands of some game masters, they can be as elaborate as a short story.

For instance, in a LARP based on the world of Hamlet, the character sheet for Ophelia might go something like this:

> You are a beautiful but delicate young woman, and things have not been going well in your kingdom lately. For one thing, the king, whom you were very fond of, has died, and his wife, Gertrude, who has been a second mother to you since your own mother died, has married his brother very quickly. This seems to have upset your boyfriend, Prince Hamlet, who was very attentive to you before his father died but has been moping around ever since. Thank goodness your brother Laertes is on his way home. He always seems to understand you. And he will divert some of the attention of your dad, who is an old dear but can be so long-winded and bossy and is always nudging you to get back together with Hamlet. He keeps thinking of embarrassing things you should do to throw yourself at Hamlet, which drives you crazy since you are very obedient but you are too modest to enjoy flirting with someone who keeps rejecting you. If only Hamlet would return to his old self.

Such a character sheet would provide the player with ideas on how to act—docile and modest and lovesick—and guide her in how to relate to other characters. It would work as a kind of "smart costume," a ready-made set of behaviors to slip into that do not require much invention to sustain but that offer opportunities for elaboration if the player is so inclined.

In addition, the character might be given a set of small sealed envelopes, or "packets," marked with instructions on when to open them. Often these are "memory packets," things a character is not to remember until an appropriate time in the game. For instance, running into Rosencrantz and Guildenstern might remind the Ophelia character of an occasion when Hamlet was particularly loving to her just before he left for college. Or she might have a packet meant to be opened after drinking a special kind of tea, a packet that might tell her that her infatuation is over and that she is now passionate about botany and has forgotten all about Hamlet. In this world, of course, her fate would be an open matter. Somewhere in her stack of packets might be written an instruction to go insane. Perhaps it would be triggered by the phrase "Get thee to a nunnery." But this would be only one of many possible paths her life could take.

In order to participate with focus in the immersive world, a character is usually given some goals to try to accomplish. For the Ophelia character, a major goal might be marrying Hamlet, and a minor one might be helping her brother get more money from their father. She would also need some hints about specific tasks that might help her achieve these goals. For instance, the overall design of the game might include a town witch and a meddlesome friar who each have potions that could affect Hamlet's behavior. Ophelia's character sheet might mention a rumor that the innkeeper knows where to get such potions. Ophelia could then set about finding out more about them, choosing which one would work, and locating and negotiating with the seller. Engaging in these activities could have repercussions for her relationship with her father. She might have to hide these activities from him or sneak off to look for them without arousing his suspicions.

A good character sheet provides a number of different plots for the player to get involved with, and a good game design would cue the various characters on how to relate to one another. The Polonius character would be told how anxious he was to make this sidetracked marriage happen. The town witch might be told to try enhancing her reputation by acquiring as clients important people who need a good herbal cure but to beware certain neighbors who will have her arrested if she is seen peddling her wares.

The person who plays Ophelia (like all the other players in the LARP) is thus supported by a world full of characters programmed to fit into her own character's plot, characters whose own intricate activities, even those that are completely unrelated to Ophelia's goals, add depth and variety to her world. The well-defined roles provide the means for each individual participant to actively create belief in the illusory world, and for all of them together to form a powerful circle of enchantment.

Regulating Arousal

According to Winnicott, "the pleasurable element in playing carries with it the implication that the instinctual arousal is not excessive"; that is, the objects of the imaginary world should not be too enticing, scary, or real lest the immersive trance be broken. This is true in any medium. If a horror movie is too frightening, we cover our eyes or turn away from the screen. If a romantic movie is too directly arousing, audience members may start necking instead of watching the characters. In the case of child's play, according to Winnicott, "instinctual arousal beyond a certain point must lead to: (i) climax; (ii) failed climax and a sense of mental confusion and physical discomfort that only time can mend; or (iii) alternate climax (as in provocation of parental or social reaction, anger, etc.)."[16] Similarly, if a participatory immersive experience is not to be pornographic and if it is not to lead to frustration or to inappropriate explosion (like the verbal tirades, or flaming, in MUDs), then the participant's arousal must be carefully regulated. The trance should be made deeper and deeper without the emotions becoming hotter and hotter.

Traditional narratives have clear conventions for regulating arousal so that it is strong enough to make the story compelling but not so strong as to render the viewer uncomfortable. Consider, for example, the filmic conventions used in the barn scene in the movie *Witness* (1985) between the Philadelphia policeman John Book (Harrison Ford) and the Amish woman Rachel (Kelly McGillis), one of the most romantic scenes in recent films. Not only are the characters attractive, but their love is forbidden (since they belong to such different cultures) and goes unconsummated throughout the movie. In this scene they are sitting together in Book's car, which is hidden in the barn, and he is fixing something on the dashboard while she holds a lantern. The radio suddenly comes on, and it is playing Sam Cooke's "Wonderful World." The scene takes them out of the car and into a shy but exuberant dance. The moment at which they decide to dance is exquisitely staged. Book, moved by the nostalgic music, backs out of the passenger side of the car while the camera follows him from just behind the driver's side. He is facing the camera across the roof of the car and tapping on the roof to the beat of the song. The moment is fraught with desire, with Book's unspoken invitation to Rachel. Then Rachel is seen moving up into the frame, her back to the camera, and he smiles at her. The seduction is addressed both to the character and to the audience. In fact, in the first moment, before Rachel gets out of the car, it is aimed almost explicitly at *us*. But Harrison Ford is not looking directly into the camera, he is looking a little to the side.

This over-the-shoulder position of the camera is a standard film technique that keeps us identified with the characters while also distanced enough so that we are reminded of the presence of the other actor in the frame of the movie and of our own exclusion from it. This combination of tremendous immediacy with a clearly demarcated border maximizes our immersion in the dramatic action.

In the café scene of the IMAX movie *Wings of Courage* (discussed in chapter 2), there is a similar moment when Val Kilmer, playing the gallant pilot Jean Mermoz, gets up to dance. He has the same movie star attractiveness that Harrison Ford has in *Witness*, and, just as Huxley warned us, the three-dimensional display makes him appear extremely present before us, much more so than on a conventional movie screen. Sitting in the theater with the 3-D goggles on, I felt myself begin to blush, as if I were actually meeting his gaze. There is a discomfort in not knowing the limits of the illusion. What if he were to come right up and ask me to dance? What if he were to extend his arms like Lord Burleigh? How far into seduction could he go without breaking the spell?

One solution to the need for boundaries and conventions in participatory narrative is to focus on exhibitionism rather than on simulated sex. Feminist critics have pointed out the pervasive use of film to linger over women's bodies. In this respect, *Witness* is unusual in that (for most of the picture) it is the male actor whose body is eroticized. When John Book takes a drink of lemonade and some if it runs down his virile neck, we see him through Rachel's eyes—as achingly attractive yet forbidden. Such a scene, in which the character is erotically displayed but made unavailable by the plot, is particularly well suited to a medium with such a riveting sense of presence. In a three-dimensional movie, the viewer is inherently placed in a situation of immobilized desire. The enticing images placed before us tease us into touching them and then evaporate in our fingers. When we have virtual reality environments with strong narrative interest, they may feel similarly poignant to us. If so, then virtual reality theaters will be a good place to stage the twenty-first-century version of the crypt scene from *Romeo and Juliet,* or any participatory story that centers on unattainable desire or tender longing for the dead. Perhaps the VR medium of the future will largely support a literature of nostalgia, full of shimmering visions of the preindustrial past.

The cyberpunk writers have offered a very different view. In Neal Stephenson's complex vision of a technological dystopia, *The Diamond Age,* "ractors," or professional interactive actors, operate avatar characters over a vast medianet, through sensors implanted in their faces and bodies. The expert ractor Miranda (named for the naïf in Shakespeare's *Tempest* who speaks of the "brave new world") takes a wide range of parts: Shakespearean heroines in role-playing adaptations (which are only pleasurable to her if done with a talented customer), a salesclerk whose image is customized to the sexual preferences of each particular customer, and even the "eternally elusive" Carmen Sandiego. Part of her job is handling the sexual improprieties of "ractiv" entertainment. For instance, while playing the role of Ilse in the ractiv equivalent of *The Mousetrap* (a long-running murder mystery set on a train in World War II Europe), a performance in which paying guests and professional ractors interact from distant locations in a shared virtual space, she is distracted by a virtual masher:

> It was nearly ruined by one of the players, who had clearly signed up exclusively for the purpose of maneuvering Ilse into bed. He turned out to be the secret SS colonel too; but he was so hell-bent on fucking Ilse that he spent the whole evening out of character. Finally Miranda lured him into the kitchen in the back of the dining car, shoved a foot-long butcher knife in his chest, and left him in the fridge. She had played this role a couple of hundred times and knew the location of every potentially lethal object on the train. (P. 108)

Miranda's very professional solution to the problem of how to deal with instinctual arousal when it threatens to disrupt the illusory world was to provide an "in character" response to inappropriately "out of character" behavior.

In live-action role-playing games, the narrative conventions that control the boundary between the real world and the illusion are called "mechanics." LARP mechanics are a kind of abstract mimicry for behaviors that would otherwise require props, danger, or physical involvement. For instance, many role-playing games represent combat by elaborate arithmetical calculations of comparative strength, force, and vulnerability values. In such a game one might see a crowd of people standing in a college corridor in the middle of the night, shouting numbers at one another, doing the math in their heads, and then turning over the name tags of those players who have been calculated to be dead. There can also be mechanics for seduction. If two characters want to have sex, the mechanic might be that they go to a place separate from other players and remain there for a certain number of minutes. They then report to the game master that they have had sex. If they want to kiss, they might just say to one another "I kiss you" and "I kiss you back."

In some ways, these mechanics are the equivalent of the fade-out technique used in movies. They signal that something is happening that can only take place place in the viewer's or interactor's imagination. The abstractly represented action can be exploited for the immersive pleasure of role-playing as, for example, when two players improvise a love scene, complete with longing looks and poignant words but no necking. Or the mechanic can be exploited for its narrative consequences. For instance, in one simulation, sex with a particular woman served as a kind of truth serum. After making love she could ask one question, which her partner had to answer truthfully. This mechanic allowed sex to be used as a game strategy independent of the players' enjoyment of the scene.

In MUDs, which are on-line role-playing environments, players have created a similar repertoire of conventions for everything from weddings to virtual pie-making. Sometimes these conventions only involve navigating through the MUD to a particular virtual room and engaging in a ritualized conversation with other MUDders. For example, I might type in "south, west, south" until the program announces, "Wedding Chapel." The program would then tell me the names of those who are present, but it would be up to all the role players together to improvise the wedding scene. In other MUDs, players can program some objects and events into the system. The Wedding Chapel could contain an automated minister, who would lead the couple through their vows. After the ceremony the minister would remember they were married; he might be programmed to tell everyone he meets about each new marriage—perhaps even gossiping about what the bride was wearing and whether she looked pregnant.

The narrative strategies used in MUDs raise many questions about how to establish boundaries between private fantasy and public enactment. There is no single storyteller in a MUD; the computer program itself serves as narrator of the story, publishing the dialogue of the players to their computer screens and announcing entrances, exits, descriptions, and some events. The command structure by which the players act in the fictional world establishes the narrative conventions. The most common conventions regulate the privacy of the dialogue: players can establish separate rooms, which function as private stages, or they can use the "whisper" command to one another, so that their conversation cannot be heard by others in the same room. If DarkBird whispers to WoodElf, "I kiss you," then the words "DarkBird whispers, 'I kiss you' " will appear only on the screens of these two players and no one else's. But if DarkBird "says" the words instead of whispering them, then everyone else in proximity to the lovers will see "DarkBird says, 'I kiss you' " on their screens. The privacy conventions allow the players to decide how much of their role-playing they want to share with the general group, but the digital stage does not always offer them complete privacy. A common grievance on MUDs is the presence of nosy wizards—the chief programmers or senior players in the virtual world who can eavesdrop on private conversations.

In other MUDs, *kiss* might be a command word; that is, if Dark-Bird types, "I kiss WoodElf" (or perhaps "kiss:WoodElf"), the system reports "DarkBird kisses WoodElf." The command convention gives the kiss the authority of a narrated event. Events that happen by command can change the state of the game (e.g., the command *Go north* changes who is where), and they can have hidden consequences. For example, if two people have virtual sex using command words, the result might be a virtual pregnancy, which would be generated by the system on the basis of a combination of random chance and the couple's virtual birth control practices. The system would then keep track of a character's pregnancy, remembering it at future sessions. It might offer an automated abortion service or provide some of the other role-playing characters with the ability to use specialized commands that allow them to perform virtual abortions or deliver virtual babies.

In some MUDs only the wizards can make up new commands; in others, all the players share this power. The issue of defining new commands becomes particularly sensitive in sexual matters. If, for example, BadTroll invents a rape command and then types in "rape: WoodElf," the system will report the action as objective reality to everyone in the room. The narration increases the victim's sense of violation. Often such events spill over into long out-of-character discussions on the social values of the virtual community. Sometimes they result in limiting the participants' ability to invent their own commands.[17]

Just as actors need conventions for staging fights and faking kisses, so too will interactors in a virtual world need specific mechanics for physical contact, mechanics that deepen the fantasy without disrupting the immersive trance. For instance, a holographic lover might offer a kiss by coming closer and then swirling away while music swells in the background. Such a "swirling" convention would emphasize the approach to the embrace and the long glance afterward rather than the kiss itself. Or an interactor wearing a special data glove might gently wave her hand, thereby signaling to her avatar figure within the frame of the virtual world that she should walk toward her lover and receive the kiss, which would be experienced through imaginative identification with the surrogate.

The computer is providing us with a new stage for the creation of participatory theater. We are gradually learning to do what actors do, to enact emotionally authentic experiences that we know are not "real." The more persuasive the sensory representation of the digital space, the more we feel that we are present in the virtual world and the wider range of actions we will seek to perform there. The ease with which MUDders and LARPers take on and cast off personas suggests that an audience is growing that has been trained in impersonation. We are all gradually becoming part of a worldwide repertory company, available to assume roles in ever more complex participatory stories. Little by little we are discovering the conventions of participation that will constitute the fourth wall of this virtual theater, the expressive gestures that will deepen and preserve the enchantment of immersion.

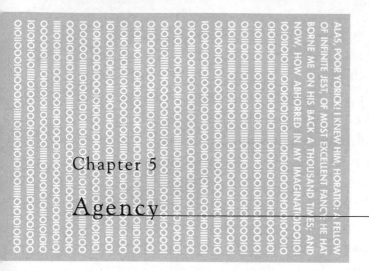

Chapter 5

Agency

The more realized the immersive environment, the more active we want to be within it. When the things we do bring tangible results, we experience the second characteristic delight of electronic environments—the sense of agency. Agency is the satisfying power to take meaningful action and see the results of our decisions and choices. We expect to feel agency on the computer when we double-click on a file and see it open before us or when we enter numbers in a spreadsheet and see the totals readjust. However, we do not usually expect to experience agency within a narrative environment.

Even in the rare circumstances when we are invited to participate in a traditional narrative form, our participation is circumscribed in a way that generally limits our sense of agency. For instance, if the audience at a performance of *Peter Pan* decided that Tinkerbell is a pest and refused to clap her back to life, the play would come to a halt. The participatory dinner theater plays that simulate an Italian wedding or an Irish wake or a Jewish funeral encourage audience participation by keeping the plot to a minimal level and the dialogue with the audience to social formulas appropriate to distant friends of the family. When the groom in such a play leans down to kiss me as a guest at his wedding, I can congratulate him and warn him about staying away from his ex-girlfriend now that he is married, but I cannot really influence his behavior. When audience members are included in the story, they serve only as the butt of a joke. They may be accused of adultery by the priest or shot by a mafioso relative. The slender story is designed to unfold in the same way no matter what individual audience members may do to join the fun.

In fact, participatory theater performances become participatory by incorporating folk art forms and festival behavior such as singing, dancing, and sharing a feast. Striking up a familiar song or dance tune—"Que Sera Sera" or a tarantella—is a reliable way to get the audience involved. Musical participatory forms are successful because they rely on careful cueing and formulaic behavior: We sing along with the chorus and remain silent for the verse; we answer the singer's "call" with the appropriate response. And we do these things in unison as a single voice. In a square dance we perform whatever steps the caller announces because the repertoire of possible movements and the rules of combination are known to both parties before the music starts, and though everyone does not have to do exactly the same thing at the same time, all the square dancers do have to be part of a single overall pattern. Folk dancing in ballroom style offers a model of freer participation. In the Cajun two-step or the Brazilian samba, for instance, all the dancers share a repertoire of movements, and each set of partners has license to invent its own combinations and interpretations of these movements. Like jazz musicians, the dancers can improvise their own satisfying creations from these given elements. But the greater individual freedom in ballroom-style folk dancing means that the group as a whole has less coherence than at a square dance.

Electronic environments have similar formulas and rules for structuring participation. For instance, when users are merely asked to respond to a menu with a predictable begin/quit choice, they are performing a kind of response to the "call" of the machine. When we learn a complicated program, like a word processor, and run through its familiar steps in order to do a difficult job, we are like participants in a square dance, repeating formulaic sequences, with the relevant manual page acting as caller of the dance. When we are placed within a simulation environment and allowed to experiment with changing a set of parameters as we see fit (more nitrogen, less algae), we are acting more like the leading partner in a Cajun dance. The crucial difference, however, between folk art rituals and computer-based interactions is that on the computer we encounter a world that is dynamically altered by our participation. On the ballroom dance floor, we can at most influence our partner, but the musicians and the rest of the dancers remain relatively unaffected. Within the world of the computer, however, when the right file opens, when our spreadsheet formulas function correctly, or when the simulated frogs flourish in the model pond, it can feel as if the entire dance hall is at our command. When things are going right on the computer, we can be both the dancer and the caller of the dance. This is the feeling of agency.

Because of the vague and pervasive use of the term *interactivity*, the pleasure of agency in electronic environments is often confused with the mere ability to move a joystick or click on a mouse. But activity alone is not agency. For instance, in a tabletop game of chance, players may be kept very busy spinning dials, moving game pieces, and exchanging money, but they may not have any true agency. The players' actions have effect, but the actions are not chosen and the effects are not related to the players' intentions. Although gamemakers sometimes mistakenly focus on the number of interactions per minute, this number is a poor indicator of the pleasure of agency afforded by a game. Some games, like chess, can have relatively few or infrequent actions but a high degree of agency, since the actions are highly autonomous, selected from a large range of possible choices, and wholly determine the course of the game.

Agency, then, goes beyond both participation and activity. As an aesthetic pleasure, as an experience to be savored for its own sake, it is offered to a limited degree in traditional art forms but is more commonly available in the structured activities we call games. Therefore, when we move narrative to the computer, we move it to a realm already shaped by the structures of games. Can we imagine a compelling narrative literature that builds on these game structures without being diminished by them? Or are we merely talking about an expensive way to rewrite *Hamlet* for the pinball machine?

The Pleasures of Navigation

One form of agency not dependent on game structure yet characteristic of digital environments is spatial navigation. The ability to move through virtual landscapes can be pleasurable in itself, independent of the content of the spaces. A friend of mine whose son is an avid *Nintendo* player complains that when he tries out the games he is annoyed at having to be fighting all the time, since the combat is an unwelcome distraction from the pleasure of moving around the unfolding spaces of the maze. For my friend, videogames are about exploring an infinitely expandable space. Similarly, new explorers of the World Wide Web find themselves entranced with the ability to leap around the world, following links from one home page or Web site to the next mostly for the pleasure of the repeated arrivals. The navigational pleasures are richly exploited by the many forms of labyrinths, from Zork-like dungeons to informational webs, that fill cyberspace. All of them allow us to experience pleasures specific to intentional navigation: orienting ourselves by landmarks, mapping a space mentally to match our experience, and admiring the juxtapositions and changes in perspective that derive from moving through an intricate environment.

This participatory pleasure is not unlike the enjoyment people find in the organized sport of "orienteering," where players follow a series of geographical clues across a large and complex terrain, such as a portion of the Maine woods or downtown Boston. Construing space and moving through it in an exploratory way (when done for its own sake and not in order to find the dentist's office or the right airport gate) is a satisfying activity regardless of whether the space is real or virtual. Electronic environments offer the pleasure of orienteering in two very different configurations, each of which carries its own narrative power: the solvable maze and the tangled rhizome.

The Story in the Maze

Zork-like puzzle dungeons and maze-based combat videogames derive from a heroic narrative of adventure whose roots are in antiquity. It was the mythical King Daedalus of Crete who built a labyrinth around the deadly Minotaur. The horrible beast required the yearly sacrifice of Athenian youths and maidens, whom it devoured, until the hero Theseus arrived to slay it. In the story, Ariadne, the daughter of the king, fell in love with Theseus and gave him a sword to kill the beast and a thread to find his way out again. Theseus's maze was therefore a frightening place, full of danger and bafflement, but successful navigation of it led to great rewards.

The adventure maze embodies a classic fairy-tale narrative of danger and salvation. Its lasting appeal as both a story and a game pattern derives from the melding of a cognitive problem (finding the path) with an emotionally symbolic pattern (facing what is frightening and unknown). The maze story celebrates the combination of intelligence and courage, and it depicts romantic love as the element that provides the hope that brings the hero into the confrontation and back out again to safety. Like all fairy tales, the maze adventure is a story about survival. The maze is a road map for telling this story.

As a format for electronic narrative, the maze is a more active version of the immersive visit (as described in chapter 4). Maze-based stories take away the moving platform and turn the passively observant visitor into a protagonist who must find his or her own way through the fun house. A typical maze-based puzzle game sends you, the player, through a multitiered space vaguely resembling an *Arabian Nights* palace. You operate an avatar who walks through the palace rooms, whose tiled floors and ornately decorated corners often hide treasures that are tricky to perceive. The palace is full of informants, who speak in text bubbles and whom you reply to from menus, and you must negotiate with them carefully, offering them icons representing money or other valuables. A mysterious peddler on one of the lower levels holds a talisman needed to get into the highest chamber. You must have it with you while you stand on a special spot that is hidden in the patterning of the floor. If you forget to get it, you must retrace your steps through many perils. The game is like a treasure hunt in which a chain of discoveries acts as a kind of Ariadne's thread to lead you through the maze to the treasure at the center.

This kind of narrative structure need not be limited to such simplistic content or to an explicitly mazelike interface. In the right hands a maze story could be a melodramatic adventure with complex social subtexts. For instance, instead of a fairy tale palace it could be set in a Kafkaesque city where the secret police are rounding up and deporting citizens with the wrong kind of papers. The protagonist's role would be to save them, a task that would require navigation through the corridors of power and through underground hiding places, elaborately conducted negotiations, clever manipulation of bureaucrats, and split-second timing. The characters in the menacing world could be subtly portrayed, in either graphics with text bubbles or in video segments. Saving people might involve horrifying choices, perhaps implicating the protagonist in the corruption of the violent world. The maze could be composed not only of spatial twists but of moral and psychological choices. Just as it is hard to see where a tangle of virtual corridors is leading, so too would it be hard to foresee the consequences of your actions and to determine what to value and whom to trust. Just as Kafka used the conventions of the fable to convey the profound depersonalization of modern life and Art Spiegelman used the format of the comic book to tell the story of his father's Holocaust experiences, a digital artist might use the structure of the adventure maze to embody a moral individual's confrontation with state-sanctioned violence.

Whether an adventure maze is simple or complex, it is particularly suited to the digital environment because the story is tied to the navigation of space. As I move forward, I feel a sense of powerfulness, of significant action, that is tied to my pleasure in the unfolding story. In an adventure game this pleasure also feels like winning. But in a narrative experience not structured as a win–lose contest the movement forward has the feeling of enacting a meaningful experience both consciously chosen and surprising. However, there is a drawback to the maze orientation: it moves the interactor toward a single solution, toward finding the one way out. The desire for agency in digital environments makes us impatient when our options are so limited. We want an open road with wide latitude to explore and more than one way to get somewhere. We want the "pullulating" web that Borges described, constantly bifurcating, with every branch deeply explorable.

Rapture of the Rhizome

The second kind of digital labyrinth, which has arisen from the academic literary community, is the postmodern hypertext narrative described in chapter 2. Full of wordplay and indeterminate events, these labyrinths derive not from Greek rationalism but from poststructuralist literary theory and are unheroic and solutionless. Like a set of index cards that have been scattered on the floor and then connected with multiple segments of tangled twine, they offer no end point and no way out. Their aesthetic vision is often identified with philosopher Gilles Deleuze's "rhizome," a tuber root system in which any point may be connected to any other point.[1] Deleuze used the rhizome root system as a model of connectivity in systems of ideas; critics have applied this notion to allusive text systems that are not linear like a book but boundaryless and without closure. Stuart Moulthrop, a theorist and electronic fiction writer, states it this way:

> Seen from the viewpoint of textual theory, hypertext systems appear
> as the practical implementation of a conceptual movement that . . .

rejects authoritarian, "logocentric" [i.e., truth-affirming] hierarchies
of language, whose modes of operation are linear and deductive, and
seeks instead systems of discourse that admit a plurality of meanings
where the operative modes are hypothesis and interpretive play.[2]

The postmodern hypertext tradition celebrates the indeterminate
text as a liberation from the tyranny of the author and an affirmation
of the reader's freedom of interpretation. But the navigational software designed specifically for this purpose and celebrated by many
proponents of literary hypertext is anything but empowering to the
reader, even in comparison to the earliest Web browsers.[3] For instance, it offers the navigating reader no way to mark links as having
been already taken, and no way to mark a lexia so it can be easily
jumped back to. Many of the stories written in this framework do not
even mark which words are hot links within the lexia text. Instead,
the reader has to click on a pop-up display of cryptic link names.
Moulthrop's own *Victory Garden,* which is perhaps the most coher-
ently structured literary hypertext, contains a clever overview map of
the major story clusters, which are arranged like a Borgesian garden
labyrinth. But readers cannot easily return to the overview in order to
get a sense of where they are or how much is left to read. In trying to
create texts that do not "privilege" any one order of reading or interpretive framework, the postmodernists are privileging confusion itself. The indeterminate structure of these hypertexts frustrates our
desire for narrational agency, for using the act of navigation to unfold
a story that flows from our own meaningful choices.

But the unsolvable maze does hold promise as an expressive structure. Walking through a rhizome one enacts a story of wandering, of
being enticed in conflicting directions, of remaining always open to
surprise, of feeling helpless to orient oneself or to find an exit, but the
story is also oddly reassuring. In the rhizome, one is constantly threatened but also continuously enclosed. The fact that the plot will not
resolve means that no irreparable loss will be suffered. The narrator
of *Afternoon* (discussed in chapter 2) will not have to confront the

fact of the morning's fatal accident so long as the afternoon's evasive wanderings continue, and the reader of *Victory Garden* does not have to accept the death of an appealing character so long as there are multiple paths to explore, including some that lead to alternate realities in which she comes back home from the war. In both stories the reader is protected from feeling the irreversibility of death by the fact that the stories do not have to end there.

The boundlessness of the rhizome experience is crucial to its comforting side. In this it is as much of a game as the adventure maze. In fact, it reminds me of a particular game my son William invented at about age five. At his own initiative he one day drew a large game board, assembled dice and playing pieces, and invited his father to join him in an inventively improvised game with ever-changing and ever more elaborate rules. After two hours of this surreal activity, my husband became restless and began asking every five minutes or so if the game was almost over. William responded by calmly walking into the kitchen, where I was sitting, and asking me to write his father the following note:

DEAR DAD—THIS GAME WILL NEVER END. WILLIAM

The rhizome has the same message. As we navigate its tangled, anxiety-laden paths, enclosed within its shape-fitting borders, we are both the exasperated parent longing for closure and separation and the enthralled child, lingering forever in an unfolding process that is deeply comforting because it can never end.

Giving Shape to Anxiety

Both the overdetermined form of the single-path maze adventure and the underdetermined form of rhizome fiction work against the interactor's pleasure in navigation. The potential of the labyrinth as a participatory narrative form would seem to lie somewhere between the two, in stories that are goal driven enough to guide navigation but open-ended enough to allow free exploration and that display a satisfying dramatic structure no matter how the interactor chooses to traverse the space.

The key to creating an expressive fictional labyrinth is arousing and regulating the anxiety intrinsic to the form by harnessing it to the act of navigation. Suspense, fear of abandonment, fear of lurking attackers, and fear of loss of self in the undifferentiated mass are part of the emotional landscape of the shimmering web. Moving through the space can therefore feel like an enactment of courage and perseverance, like Gary Cooper's striding through the town in *High Noon*. Computer gamers often experience shivers of physical fear as they approach an unopened door in a text-based or graphics-based labyrinth. The drama of suspenseful approach does not have to be tied to combat or to jack-in-the-box effects. It can also have the feeling of a determination to face the truth, to stare directly at the threatening beast. It can be experienced by the navigating reader/viewer as well as by the player/protagonist.

One such use of the labyrinth as a means of evoking and controlling terror is a story format increasingly used by my fiction students, a story I have come to call the "violence hub." Writers will place an account of a violent incident, often a real or invented newspaper article, at the center of a web of narratives that explore it from multiple points of view. A helicopter accident near MIT, a robbery in a convenience store, a canoeing fatality—these are all examples of an act of sudden violence that have served as the center of such a story web. The incident account itself is usually bare but evocative. People have died in violent and surprising circumstances, we are told. The incident happened at a particular place and time involving a particular group of people. The names in the account or in a diagram of the site of the accident lead outward with hot links to the satellite files that tell us how the incident appeared to the various people involved— the perpetrator, the witnesses, the rescuers, the victims, the survivors.

The proliferation of interconnected files is an attempt to answer the perennial and ultimately unanswerable question of why this incident happened. For instance, one convenience store robbery labyrinth follows the robber, the clerk, the owner, and the cop (who shoots the robber) back through the events leading up to the shoot-out and forward again into the moment of violence.[4] Reading it we feel sympathy with all of them, and we see how they appeared to one another. A web story of a teenager who drowned on a white-water canoeing trip describes the traumatic experience from the points of view of the friends he was traveling with, the operators of the canoeing company, the emergency medics, and the family members receiving the dreadful phone call. These violence-hub stories do not have a single solution like the adventure maze or a refusal of resolution like the postmodern stories; instead, they combine a clear sense of story structure with a multiplicity of meaningful plots. The navigation of the labyrinth is like pacing the floor; a physical manifestation of the effort to come to terms with the trauma, it represents the mind's repeated efforts to keep returning to a shocking event in an effort to absorb it and, finally, get past it. The retracing of the situation from different perspectives leads to a continual deepening in the reader's understanding of what has happened, a deepening that can bring a sense of resolution but one that allows for the complexity of the situation and that leaves the moment of shock unchanged and still central.

A linear story, no matter how complex, moves toward a single encompassing version of a complex human event. Even those multiform stories that offer multiple retellings of the same event often resolve into a single "true" version—the viewpoint of the uninvolved eyewitness or the actual reality the protagonists wind up in after the alternate realities have collapsed. A linear story has to end in some one place: the last shot of a movie is never a split screen. But a multi-threaded story can offer many voices at once without giving any one of them the last word. This is a reassuring format for encountering a traumatic event because it allows plenty of room for conflicting emotions. It lets us disperse complex, intense reactions into many derivative streams so that we do not have to feel the full flood of sorrow all at once. The multithreaded web story achieves coherent dramatic form by shaping our terror into a pattern of exploration and discovery.

The Journey Story and the Pleasure of Problem Solving

The navigational space of the computer also makes it particularly suitable for journey stories, which are related to mazes but offer additional opportunities for exercising agency. Journey stories date back to oral storytelling, from the fairy-tale convention of setting out from home to find one's fortune to the voyages of Odysseus and Sinbad. It is a universal archetype recognizable across all the variations of culture, author, and medium. After the invention of the printing press, the journey story was reinvented as the picaresque novel, exemplified by *Don Quixote, Moll Flanders, Tom Jones, Huckleberry Finn,* and *Catcher in the Rye.* With the invention of the movie camera, the journey story was again revived, and its variations include everything from the comic (e.g., Buster Keaton's *The General*) to the cowboy melodrama (e.g., *Stagecoach, The Searchers*) and the feminist buddy film (*Thelma and Louise*). When television came along, journey stories (*Wagon Train, Route 66, The Fugitive,* and, of course, *Star Trek*) were among the most successful series.

Moving the journey story from the fairy tale to the novel meant moving it from a symbolic realm of universal actors (a king, a wicked stepmother) to a particularized social world and a particular time and place. In the novel the cruel things that happen to the hero are often treated as instances of a specific social injustice, like the English Poor Laws, rather than as the work of a generic antagonist like a big bad wolf. Moving the journey to the movies opened up the visual dimension of the archetype. Journey films often emphasize exotic landscapes, foreign cultures, and the lure of open spaces. Since television is best at portraying interior dramas and family-size social units,[5] journey stories on TV generally focus on a succession of small communities or even replace the hero and sidekick with an entire traveling community, as on *Star Trek*. On the computer the journey story emphasizes navigation—the transitions between different places, the arrivals and departures—and the how-to's of the hero's repeated escapes from danger.

One of the consistent pleasures of the journey story in every time and every medium is the unfolding of solutions to seemingly impossible situations. We watch each new situation along the road and wonder how the hero will escape a beating or a hanging or a forced marriage or jailing. When Odysseus foolishly allows himself to be captured by the Cyclops, a huge, one-eyed man-eater, he is presented with a life-and-death riddle. The situation is carefully described so that it seems that he has no chance of getting out. The Cyclops is a brutal and heartless creature who brags that he is unafraid of Zeus and therefore free to do what he pleases. Every night two more of Odysseus's men are eaten; the survivors know they must get out soon or die. They could kill the monster as he sleeps, but the cave is barred with a stone too heavy for them to move; if they kill him, they will never get out again. The Cyclops's routine is unvaried: he goes off with the sheep in the morning, closing the cave behind him, and comes back at night with the sheep, ready for a dinner of Greek sailors. Then Odysseus (who is narrating the story) tells us how he solves the problem. He prepares some wine. He prepares a battering ram and gathers a group of strong helpers. He tells the giant his name is "Nobody" and gets him very drunk. When the Cyclops falls asleep, Odysseus and his men heat the battering ram in the fire and thrust it into his one terrible eye. Now the giant is blinded, but how will the Greeks get out? While the Cyclops is raging, Odysseus separates the sheep into groups of three and places each of his men under the middle sheep and himself beneath the strongest ram in the flock. Finally, the Cyclops lets out his sheep, as Odysseus has seen him do every morning, and out go all the Greeks with them. And when the Cyclops complains about his tormentor to Zeus, how does he refer to him? He calls him by name: Nobody. Odysseus's description is constructed so that we can enjoy each individual step and gain increasing pleasure as the overall plan becomes clear. The story is as much a riddle as Oedipus's, but the answer to the riddle is not in a single word; it is in a series of beautifully orchestrated steps, an elegant algorithm for defeating giants.

Computer-based journey stories offer a new way of savoring exactly this pleasure, a pleasure that is intensified by uniting the problem solving with the active process of navigation. On the computer the dramatic situation of capture and escape can be simulated by keeping the player within a confined space until the solution to a puzzle is found. These puzzles are most satisfying when the actions have a dramatic appropriateness, when they serve as a way of increasing our belief in the solidity and consistency of the illusory world. For instance, in *Myst* the wizard's island includes an elevator hidden in a giant tree and operated by a nearby control panel. In addition to solving the puzzle of the panel, the interactor must move efficiently through the space to get to the elevator at just the right moment in its descent. The concreteness with which the space is detailed makes the sequence feel not like a test of coordination but like a dramatic moment. By contrast, in the computer game *The Seventh Guest*, the player is asked to cut up a cake into enough segments to match the number of murder victims. The puzzle is a satisfying one, but since there is no one there to eat the cake, the action takes us outside the immersive world instead of reinforcing our belief in it.

The most dramatically satisfying puzzles are those that encourage the interactor to apply real-world thinking to the virtual world. For instance, a computationally sophisticated MIT student who is also an expert gamer instanced a particular dramatic moment from the text-based *Zork II* as among his lifetime favorites: The story involves a dragon that is slow to rouse but always lethal if you persist in fighting him. Elsewhere in the dungeon is a wall of ice that is impossible to pass. What you must do is attack the dragon enough to get his attention—but not so much that he "toasts" you—and then run and head for the wall of ice. The dragon follows, sees his reflection in the ice, and thinks it is another dragon. He rears up and breathes fire at his enemy; as he does so, the ice melts, drowning the dragon and eliminating the obstructing wall.[6] Like Odysseus in the Cyclops's cave, the player escapes by outsmarting a ferocious monster using only the materials at hand.

Games into Stories

Games seem on the face of it to be very different from stories and to offer opposing satisfactions. Stories do not require us to do anything except to pay attention as they are told.[7] Games always involve some kind of activity and are often focused on the mastery of skills, whether the skill involves chess strategy or joystick twitching. Games generally use language only instrumentally ("checkmate," "ball four") rather than to convey subtleties of description or to communicate complex emotions. They offer a schematized and purposely reductive vision of the world. Most of all, games are goal directed and structured around turn taking and keeping score. All of this would seem to have nothing to do with stories.

In fact, narrative satisfaction can be directly opposed to game satisfaction, as the endings of *Myst*, widely hailed as the most artistically successful story puzzle of the early 1990s, make clear. The premise of the *Myst* story is the confinement of two brothers, Sirrus and Achenar, in magical books that serve as a dungeon. Through a video window we can see them in their imprisonment and hear them talking to us in short, staticky segments. Each one warns us about the wickedness of the other and asks us to rescue him. The brothers can only be freed by heroic labors of problem solving by the player, who must journey to four magical lands or ages and bring back a single page from each of them for either Sirrus or Achenar. Each time the player gives one of the brothers a magic page, he responds with a slightly clearer video segment. At the end of the game, when most of the puzzles have been solved, the player has most likely gone to each land twice in order to gather both sets of pages and to hear all of the messages from both brothers. At this point we are faced with a dramatic choice. The last magic page will release one or the other of them from the book. Which is it to be?

The game is well designed in that all the evidence on which to base a decision is, as in any good detective story, available to the player. Exploring the various lands reveals—through accusatory notes, hidden corpses, imperial furnishings, desolated landscapes, and multiple instruments of torture and destruction—the villainy of both brothers. The secret of the game is that although both brothers are evil, their father, Atrus, is alive and—with some more puzzle solving—can be found and rescued. The "winning" ending involves locating the good wizard Atrus and remembering to bring with you the magical item that will free him from captivity. This is a satisfyingly fair yet challenging mystery plot.

Yet surprisingly, the "losing" endings of the game are much more satisfying than the winning ending. In the winning ending one finds a beautifully rendered but dramatically inert video cutout of Atrus superimposed on a backdrop of a very shallow fantasyland. Unlike all the other lands visited during the game, this one is not really explorable and offers no pleasures of manipulation. It is a dead end. The ending in which you get to the wizard but forget to bring him the means of escape is more dramatic, because he gets quite angry at you. But the most dramatically satisfying endings are the near-identical losing branches, which are the result of choosing to rescue either of the evil brothers. The moment you release either Achenar or Sirrus from imprisonment, he will mockingly turn on you and lock *you* in the very same dungeon from which he has escaped! The visual effect is simple but brilliantly effective because it reverses your perspective. Throughout the game you have peered into each brother's dungeon through a static-ridden, credit-card-size window embedded with the parchment page of an enchanted book. The brothers' immobility has been marked by the fact that you could see little more than their faces. Now you are looking out through a similarly staticky window set into a totally black screen. Through the window you can see the evil brother now exultantly standing and moving around while looking down at you, just as you had looked down at him. A game that marked a breakthrough in ease of navigation appropriately ends by immobilizing the player.

The superiority of the losing endings of *Myst* suggest a basic opposition between game form and narrative form. How can we tell significant stories in a form that always has to end happily? How can we impose endings that yield complex story satisfactions on a form that is based on win/lose simplicity? Many would argue that computer-based narrative will always be gamelike and that such dissatisfactions are therefore inevitable. But when looked at more closely, games and stories are not necessarily opposed.

Games as Symbolic Dramas

A game is a kind of abstract storytelling that resembles the world of common experience but compresses it in order to heighten interest.[8] Every game, electronic or otherwise, can be experienced as a symbolic drama. Whatever the content of the game itself, whatever our role within it, we are always the protagonists of the symbolic action, whose plot runs like one of the following:

- I encounter a confusing world and figure it out.
- I encounter a world in pieces and assemble it into a coherent whole.
- I take a risk and am rewarded for my courage.
- I encounter a difficult antagonist and triumph over him.
- I encounter a challenging test of skill or strategy and succeed at it.
- I start off with very little of a valuable commodity and end up with a lot of it (or I start off with a great deal of a burdensome commodity and get rid of all of it).
- I am challenged by a world of constant unpredictable emergencies, and I survive it.

Even in games in which we are at the mercy of the dice, we are still enacting a meaningful drama. Playing purely luck-based games is captivating because we are modeling our basic helplessness in the universe, our dependence on unpredictable factors, and also our sense of hopefulness. The people who line up at my neighborhood convenience store for lottery tickets can be seen either as dupes or as risk takers engaging in a playful ritual of faith in the benevolence of forces beyond their control. In fact, even when we lose, we are still part of the symbolic drama of the game. In that case the plots might go like this:

- I fail at an important test and suffer defeat.
- I decide to try again and again until I finally succeed.
- I decide to win by cheating, that is, by acting outside the rules, because authority is meant to be flouted.
- I realize that the world is rigged against me and others like me.

In games, therefore, we have a chance to enact our most basic relationship to the world—our desire to prevail over adversity, to survive our inevitable defeats, to shape our environment, to master complexity, and to make our lives fit together like the pieces of a jigsaw puzzle. Each move in a game is like a plot event in one of these simple but compelling stories. Like the religious ceremonies of passage by which we mark birth, coming of age, marriage, and death, games are ritual actions allowing us to symbolically enact the patterns that give meaning to our lives.

Games can also be read as texts that offer interpretations of experience. For instance, the board game Monopoly can be read as an interpretation of capitalism, an enactment of the allures and disappointments of a zero-sum economy in which one gets rich by impoverishing one's neighbors. Or it can be read as a patterned expression of our knowledge that success in life is always the result of both planning and chance. When we play Monopoly, we are taking part in a structured drama that offers, in addition to its win/lose ending, moments in which we give expression to our ambition, greed, and benevolence and our tendencies to take risks and exploit others. Even a game with no verbal content, like Tetris, the wildly popular and powerfully absorbing computer game of the early 1990s, has clear dramatic content. In Tetris irregularly shaped objects keep falling from the top of the screen and accumulating at the bottom. The player's goal is to guide each individual piece as it falls and position it so that it will fit together with other pieces and form a uniform row. Every time a complete row forms, it disappears. Instead of keeping what you build, as you would in a conventional jigsaw puzzle, in Tetris everything you bring to a shapely completion is swept away from you. Success means just being able to keep up with the flow. This game is a perfect enactment of the overtasked lives of Americans in the 1990s—of the constant bombardment of tasks that demand our attention and that we must somehow fit into our overcrowded schedules and clear off our desks in order to make room for the next onslaught.[9]

If the same spatial ideas behind the movement of the colored shapes in Tetris—relentless activity, misfits and tight couplings, order and chaos, crowding and clearing—are represented in a dance, we automatically associate them with ordinary human experience, because we see human beings enacting them. In the computer game the interactor is the dancer and the game designer is the choreographer. The screen objects are like a symbolic language for inducing our activity. So while we experience the game as being about skill acquisition, we are drawn to it by the implicit expressive content of the dance. Tetris allows us to symbolically experience agency over our lives. It is a kind of rain dance for the postmodern psyche, meant to allow us to enact control over things outside our power.

Games are recreational because they offer no immediate benefit to our survival. Yet game-playing skills have always been adaptive behaviors. Games traditionally offer safe practice in areas that do have practical value; they are rehearsals for life. Lion cubs roughhouse with one another in order to grow to be predators. Small children still play hide-and-seek, a good way of training hunters, and ring-around-a-rosy, a good way of practicing cooperation and coordinated behaviors. Older children in our society are understandably drawn to pitting themselves against machines. The violence and simplistic story structure of computer skill games are therefore a good place to examine the possibilities for building upon the intrinsic symbolic content of gaming to make more expressive narrative forms.

The Contest Story

The most common form of game—the agon, or contest between opponents—is also the earliest form of narrative. This is not surprising since opposition is one of the most pervasive organizing principles of human intelligence and language.[10] Just as we automatically organize the temporal and spatial world into opposing characteristics (night/day, up/down, right/left), so too do we look at the things that happen in the world in terms of struggles between opposites (God/Satan, male/female, Cain/Abel, Jews/Gentiles). The Greek word *agon* refers to both athletic contests and to dramatic conflicts, reflecting the common origin of games and theater. A simple shoot-'em-up videogame, then, belongs to the extremely broad dramatic tradition that gives us both the boxing match and the Elizabethan revenge play.

Most of the stories currently told on the computer are based on the structure of a contest of skill. The interactor is given the role of a fighter or detective of some sort and is pitted against an opponent in a win/lose situation. From their beginnings in the 1970s, computer games have developed multiple representations of the opponent, who may be another human player (as in the first videogame, *Pong*), a character embedded in the story (as in *Pacman*), and the programmer or game designer implicit in the game (as in *Zork*). Contest games have also developed at least three different ways of situating the player: we can watch from a spectator perspective while operating our own avatar character or spaceship (as in *Mortal Kombat*); watch from a situated perspective while operating a character (as in *Rebel Assault*, where we see the vehicle we are operating as if we are following just behind it with a movie camera); or, most immersively, watch and act from a situated first-person viewpoint, as in *Doom*, where we see the landscape of the game and our opponents coming toward us as if we are really present in space. These gaming conventions orient the interactor and make the action coherent. They are equivalent to a novelist's care with point of view or a director's attention to staging.

Fighting games have also developed a sure-fire way of combining agency with immersion. The most compelling aspect of the fighting game is the tight visceral match between the game controller and the screen action. A palpable click on the mouse or joystick results in an explosion. It requires very little imaginative effort to enter such a world because the sense of agency is so direct. The imaginative engagement is even stronger with an arcade-style interface that lets you sit in a brightly painted model of a spaceship or fire a toy gun. My own surprising immersion in the *Mad Dog McCree* arcade game (discussed in chapter 2) depended heavily on the heft and six-shooter shape of the laser gun controller and on the way it was placed in a hip-height holster ready for quick-draw contests. As soon as I picked up that gun, I was transported back to my childhood and to the world of TV Westerns. When my son brought home the videogame version, based on a multibutton controller, I could not get interested in the game at all (although he liked it better that way, since it was the skill mastery that interested him rather than the story). For me, the six-shooter was an ideal threshold object, a physical device I could hold in my hand that was also an imaginary device in the world of the story. I only had to put my hands around it to enter the immersive trance. Ideally, every object in a digital narrative, no matter how sophisticated the story, should offer the interactor as clear a sense of agency and as direct a connection to the immersive world as I felt in the arcade holding a six-shooter-shaped laser gun and blasting away at the outlaws in *Mad Dog McCree*.

Because guns and weaponlike interfaces offer such easy immersion and such a direct sense of agency and because violent aggression is so strong a part of human nature, shoot-'em-ups are here to stay. But that does not mean that simplistic violence is the limit of the form.

Though violent games have dominated computer entertainment sales, there are some signs of a more complex approach. In many fighting games, like *Mortal Kombat,* the player can switch sides and play through the same confrontation from opposing positions. The *Star Wars* series of computer games offers a particularly dramatic change in player position. Most of the games, like the popular *Rebel Assault* CD-ROMs, put the player in the position of a fighter in the forces led by the heroes, Luke Skywalker, Han Solo, and Princess Leia, but the *Tie Fighter* game casts the player as a member of the Empire forces. As one adult player, a pony-tailed programmer from San Francisco, told me, this recruitment into the forces of the Empire can be a source of intense fascination. "I got totally identified with the Empire and its goals of maintaining order. I found myself hating the rebels because they brought disorder. It really freaked me out. I could see right away how I could become a great fascist." Of course, it is possible to play the game purely for the thrill of flying the Empire's planes, but the moral impact of enacting an opposing role is a promising sign of the serious dramatic potential of the fighting game.

The success of the fighting contest games poses a challenge to the next generation of digital artists. The contest format is open to expressive expansion in many ways once we move the protagonist beyond the role of a simple fighting machine. We need to find substitutes for shooting off a gun that will offer the same immediacy of effect but allow for more complex and engaging story content. We need to find ways of drawing a player so deeply into the situated point of view of a character that a change of position will raise important moral questions. We need to take advantage of the symbolic drama of the contest format to create suspense and dramatic tension without focusing the interactor on skill mastery.

Constructivism

An MIT freshman recently confided to me that he was spending a lot of time on a MUD even though he was bored with the dragon slaying that formed its main focus of activity. He continued to log on because he had figured out a way to hold parties there. He no longer used the commands for moving around and for killing, carrying, and eating beasts to build up his score as a player. Instead, he had organized other members of the MUD to use these same commands to gather provisions and bring them to a common place at a prearranged time. Dragon slaying had become an electronic form of catering.

The student's ingenuity is typical of the MUD culture. He was taking the materials at hand and repurposing them for his own uses. The notion of reassembling a fixed set of materials into new expressive form was inherent in the original *Zork*, the ancestor of the MUDs, which provided the interactor with a large vocabulary of commands and a rich array of objects that could be combined in multiple ways. MUDs began as collective games of *Zork* (hence their original name, Multi-User Dungeons). But for many people, like my student, the pleasure of sharing a virtual space in which they could chat with one another over the Internet was greater than the pleasure of the game. In the late 1980s, James Aspnes, then a graduate student at Carnegie Mellon University, created a new kind of MUD that emphasized typed conversation among the interactors and offered participants access to the programming language itself.[11] Instead of playing to increase their score, MUDders now indulged in more intense roleplaying. And with the increase in immersive involvement came a desire to construct their own virtual worlds.

Since objects in a text-based MUD are made out of programming code and words, there is no limit to what can be called into being within the virtual world. An expert MUDder might have his own private castle, with hidden pathways and working drawbridges; he could recruit other people to come live in it and swear fealty to him, or he could amuse "newbie" visitors with puzzle rooms or frighten them off with ferocious trolls. Even a very uncertain programmer can create objects with personal resonance, like a Chinese dancing fan that only looks graceful in the hands of its creator. Most of all, the power to create objects procedurally (by specifying not just their appearance but their behavior) has led to an outpouring of whimsy and practical jokes: a plate of spaghetti "squirms uneasily" whenever someone says they are hungry; a bucket of water falls on people who try to enter a player's room; magic spells turn fellow players into frogs or make them invisible to one another. MUDders relish one another's ingenuity in stretching the representational powers of the environment. This constructivist pleasure is the highest form of narrative agency the medium allows, the ability to build things that display autonomous behavior.[12]

The goal of the MUDders seems to be to be able to represent every activity from real life and fantasy fiction within the virtual world. Not everyone would enjoy the fantasy content of MUDs or the role-playing activities they support, but the changing emphasis of MUDs suggests a general trend in the exercise of agency in digital environments. The current constructivist MUD culture was built by an academic community that has enjoyed twenty years of consistent access to computers. It may well be a predictor of future trends in the larger population, which is just starting to come on-line. As computer access spreads, it is likely that more and more people will turn from win/lose game playing to the collective construction of elaborate alternate worlds.[13]

Virtual reality researcher Brenda Laurel has argued that VR environments should be reserved for constructivist adult make-believe:

If . . . the goal is to create a technologically mediated environment where people can *play*—as opposed to being entertained—then VR is the best game in town. When children play, they typically use their imaginations quite actively and constructively to invent action and assign meaning to materials (or make or find new ones) as the need arises. In VR as in children's play there is no sharp distinction between "authoring" and "experiencing." With [Laurel's VR environment] Placeholder, we learned that adults can play in the same way—when their imaginations are booted up by a rich virtual environment.[14]

But Placeholder is just a demonstration environment, and its interactors are very dependent on the suggestions of a goddess figure who proposes things for them to do and actively discourages all attempts at shooting games. We have a lot more to learn before we can reliably "boot up" the adult imagination enough to provide a completely constructivist digital environment.

One essential component of such an environment would be a repertoire of expressive gestures beyond the current staples of navigation and attack movements. The graphics-based environment of *Myst* offers a wonderful range of concrete actions made real by the textured graphics and the careful sound design. But it is a completely depopulated world. The Woggles world of greetings and imitative gestures (discussed in chapter 4) suggests that designers can use movement as a social language. The most expressive gesture I have yet experienced on the computer is petting my digital dog, Buttons, who lives on my home Macintosh screen and growls and pants appreciatively as I move a hand-shaped cursor over him by rolling the mouse. Certainly we could have stories in which we rock a baby's cradle or cover a sick person with a blanket or open a door to offer shelter to someone fleeing from a mob. It may be hard to picture such gestures in the game interfaces of today, which are often no more expressive than pushing buttons on a bank machine. But there is no reason why gestures could not be animated in a way that very closely matches the visual display with the interactor's movement and heightens the dramatic impact of the story.

Such constructivist stories will probably evolve out of the current MUD environment. The MUDs now offer a wide repertoire of commands, objects, and ritualized scenes. Soon they may feature 3-D landscapes and graphical avatars with typed-in dialogue appearing in bubbles over their heads. These developments could make it easier for a wider audience to participate in collective fantasy.

But collective fantasy can be fraught with problems. MUDders tend to fight with one another both in and out of character. They resent the power of the wizards and gods who can eavesdrop, reassign treasure, and kill or revive players. They have difficulty settling disputes over when it is acceptable to kill another player or who is entitled to the treasure left on the virtual corpses of dead adventurers. Because of the improvised nature of MUDding, a lot of time is spent in negotiating appropriate behavior rather than in story making. MUDders often tell me how much they enjoy being in character and performing the routine actions of the parts they play (recruiting squires, negotiating treaties, casting spells), but they also complain that a good MUD story is hard to sustain. They miss the sense of drama they enjoy in the fantasy literature that inspired these on-line fantasy worlds.

Perhaps the most successful model for combining player agency with narrative coherence is a well-run LARP game. Live-action role-playing games are guided by a clear aesthetics that divides plot responsibility between the game master (GM) and the players. The GM is responsible for inventing an enticing world with many things to do in it, a world populated by clearly drawn characters and offering a good dramatic mix of challenges and surprises. In a successful game the players have a great deal of constructive freedom in improvising the story and multiple ways of accomplishing their goals. If a player wants his or her character to take an action that will change the plot tremendously (say, for example, that a player wants her character to poison her husband, who also happens to be the head of the rebel army), the GM cannot prevent the player from proceeding merely because the action was unforeseen. But if the GM were to introduce a sudden hurricane or a nonplayer character in the middle of an ongoing game in order to enhance the plot, this would be considered unfair. The rule of successful game mastering is to set the world in motion, or wind up the clock, and then step back and let the plot unfold at the will of the players. However, part of what keeps live-action games cooperative is the fact that people interact face-to-face and often have continuing relationships with one another beyond the events of a game session.

Computer-based role-playing stories aim for the same degree of player freedom as the LARPs, but they often depend upon the ongoing intervention of the MUD "wizards" to avoid lapsing into plotless socializing or repetitive vignettes. There is a growing demand among MUDders for computer-based games that will maximize both dramatic structure and player freedom. Producing such systems will require the union of computer science expertise with participatory storytelling artistry. Perhaps the next Shakespeare of this world will be a great live-action role-playing GM who is also an expert computer scientist.

The Interactor as Author

One of the key questions that the practice of narrative agency evokes is, To what degree are we authors of the work we are experiencing? Some have argued (with either elation or horror) that an interactor in a digital story—not just the improvising MUDder, but even the navigating reader of a postmodern hypertext—is the author of the story. This is a misleading assertion. There is a distinction between playing a creative role within an authored environment and having authorship of the environment itself. Certainly interactors can create aspects of digital stories in all these formats, with the greatest degree of creative authorship being over those environments that reflect the least amount of prescripting. But interactors can only act within the possibilities that have been established by the writing and programming. They may build simulated cities, try out combat strategies, trace a unique path through a labyrinthine web, or even prevent a murder, but unless the imaginary world is nothing more than a costume trunk of empty avatars, all of the interactor's possible performances will have been called into being by the originating author.

Authorship in electronic media is procedural. Procedural authorship means writing the rules by which the texts appear as well as writing the texts themselves. It means writing the rules for the interactor's involvement, that is, the conditions under which things will happen in response to the participant's actions. It means establishing the properties of the objects and potential objects in the virtual world and the formulas for how they will relate to one another. The procedural author creates not just a set of scenes but a world of narrative possibilities.

In electronic narrative the procedural author is like a choreographer who supplies the rhythms, the context, and the set of steps that will be performed. The interactor, whether as navigator, protagonist, explorer, or builder, makes use of this repertoire of possible steps and rhythms to improvise a particular dance among the many, many possible dances the author has enabled. We could perhaps say that the interactor is the author of a particular performance within an electronic story system, or the architect of a particular part of the virtual world, but we must distinguish this derivative authorship from the originating authorship of the system itself.

Interestingly enough, the question of authorship in formulaic media is one that students of ancient oral narrative have considered at length. In the 1930s, Greek scholars were distressed when literary analysis revealed that Homer (and other epic preliterate poets) created through a process that involved fitting stock phrases and formulaic narrative units together. Critics at that time resisted the thought that the great artist Homer was not original in the same way that modern print-based writers are expected to be. Now, with the advent of computer-based authorship, we are experiencing the opposite confusion. Contemporary critics are attributing authorship to interactors because they do not understand the procedural basis of electronic composition. The interactor is not the author of the digital narrative, although the interactor can experience one of the most exciting aspects of artistic creation—the thrill of exerting power over enticing and plastic materials. This is not authorship but agency.

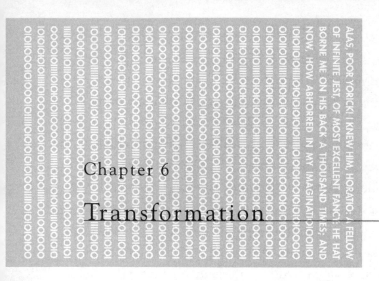

Chapter 6

Transformation

The third characteristic pleasure of digital environments is the pleasure of transformation. Computers offer us countless ways of shape-shifting. Using "morphing" software, we can transform faces so seamlessly that a grinning teenage boy melts into a haggard old woman, as if under a magic spell. Putting on a virtual reality helmet, we earthbound interactors find ourselves transmuted into soaring crows. The computer captures processes, and it therefore is always suggesting processes to us even when it is just displaying information. Anything we see in digital format—words, numbers, images, moving pictures—becomes more plastic, more inviting of change.

The transformative power of the computer is particularly seductive in narrative environments. It makes us eager for masquerade, eager to pick up the joystick and become a cowboy or a space fighter, eager to log on to the MUD and become ElfGirl or BlackDagger. Because digital objects can have multiple instantiations, they call forth our delight in variety itself. The digital Barbie doll has fifteen hundred possible outfits. A fight in *Mortal Kombat* can be between any two of sixteen distinct opponents, with the interactor taking either side of any combat. MUDs compete with one another over the variety of roles they support (elves, hobbits, Magi, thieves, enchanted animals, etc., each category having its own possible powers and abilities). In computer games we do not settle for one life, or even for one civilization; when things go wrong or when we just want a different version of the same experience, we go back for a replay.

As the medium matures, this same love of variation can be exploited for more subtle effects. Borges anticipated the pleasures of a multivariant world when he offered this glimpse of two parallel scenes in the labyrinthine novel imagined in "Forking Paths":

> In the first, an army marches into battle over a desolate mountain pass. The bleak and somber aspect of the rocky landscape made the soldiers feel that life itself was of little value, and so they won the battle easily. In the second, the same army passes through a palace where a banquet is in progress. The splendor of the feast remained a memory throughout the glorious battle, and so victory followed. (P. 98)

The juxtaposition of these two different experiences of the same winning battle reminds the reader that life itself is simultaneously desolate and splendid. By varying the texture of their experience in this way, Borges makes the doubly winning soldiers seem all the more vulnerable.

But such constant shape-shifting is also disconcerting. How can a writer tell a connected story in so fluid an environment? How will the interactor know when it begins and ends? Just as we need to define new narrative conventions for entering the immersive world and for exercising agency within it, so too do we need a new set of formal conventions for handling mutability. These conventions will arise as we get a clearer understanding of what kinds of pleasures we will seek from a literature of transformation.

Kaleidoscopic Narrative

One way to understand the new narrative environment is through the metaphor of the kaleidoscope. As Marshall McLuhan pointed out, the communications media of the twentieth century are mosaic rather than linear in structure, as compared to the printed book. Newspapers are made up of many stories calling for our attention on a single page, films are mosaics of individual shots, and television is even more mosaic in the age of the remote control than it was when McLuhan wrote about it. These mosaic informational formats have created mosaic patterns of thought that we now take for granted. We are now used to viewing the front page of a newspaper without being overwhelmed, because we have learned to take in multiple kinds of information in one quick snapshot. Similarly, years of viewing films have allowed us to automatically assemble their discontinuous images into larger patterns of continuity. We can do this because we know how to read the conventions of these media. We are guided by headline size and story placement to find our way among the many different kinds of stories in the newspaper. We know how to construct continuous space in a film by matching an exterior shot with an interior shot, and we know that a change in lighting or focus signals a flashback or a subjective memory. Mosaic organization is valuable because it gives us the overview of the front page and the faster narrative pace of the movie. We also savor the juxtapositions that these mosaic forms make possible: the shots of the Mafia godfather at the baptism intercut with shots of the murders he has ordered, the liberal op-ed column side by side with the conservative op-ed column, the cheery 1950s sitcom one click of the remote away from the lurid 1990s talk show.

The computer presents us with the spatial mosaic of the newspaper page, the temporal mosaic of film, and the participatory mosaic of TV remote control. But even while it combines the confusing multiplicity of these mosaic media, the computer offers us new ways of mastering fragmentation. It gives us "search engines" and ways to "tag" the fragments so that we can find things that are related to one another. It can preserve the history of our particular path through a web so that we can retrace it. It can generate Web pages on the fly that display exactly the kinds of things we want we to see. It offers us

a multidimensional kaleidoscope with which to rearrange the fragments over and over again, and it allows us to shift back and forth between alternate patterns of mosaic organization.

This kaleidoscopic structure has many possibilities for narrative, but
one of the most compelling is the ability to present simultaneous actions in multiple ways. In a novel, simultaneous actions are presented
consecutively. Whether we are taken to opposing camps on a battlefield or into different minds around a dining table, the simultaneous
events are often described in overlapping rather than in completely
parallel time segments; the action of the story keeps moving forward as
our point of view shifts. In a film or TV show, we can cut back and
forth between several simultaneous events during a brief sequence (as
in the baptism–murder sequence in *The Godfather*) or between two or
three narrative threads over a longer time period (as in different
patient crises in various parts of the hospital on *ER*), but the longer
the segments, the more likely it is that the time of the story is advanc-
ing with each scene rather than being retraced in different locales.
On the computer we can lay out all the simultaneous actions in
one grid and then allow the interactor to navigate among them. We
can have the expansiveness of the novel with the rapid intercutting of
the film.

How do we do this without overwhelming the interactor? We will
need a coherent set of conventions for signalling interactors when
they can move from one simultaneous action to another and for helping them keep track of where and when the various actions are taking
place. We can start by building upon the theatrical conventions of
exits and entrances, but in a digital story the stage of the narrative
may have multiple sets, that is, many separate places that are within
the scope of the dramatic action. Whenever characters move among
these sets, the reader/interactor should be able to follow them, in just
the way a camera might follow an actor from room to room. The story
line of such a multistage narrative will have to be structured in such a
way that it arouses readers' curiosity, enticing them from one set to
the next.

There have already been a few live theatrical presentations that have moved toward such a dramatic structure. One concrete model for such a story is the comical trilogy by Alan Ayckbourn, collectively called *The Norman Conquests,* which was originally performed in London and New York on three successive nights.[1] The trilogy revolves around three couples: Annie, an unmarried woman who is taking care of her cranky bedridden mother, and her overly shy suitor, Tom, a veterinarian; Ruth, Annie's businesslike sister, and her womanizing husband, Norman; and Reg, brother to Annie and Ruth, and his overcontrolling wife, Sarah. The farcical action takes place in Annie's house and involves Norman's flirtation with both Annie and Sarah while trying to placate his wife, Ruth; Tom's slow movement toward courtship with Annie; and everyone's quarrels with everyone else. This inventive romantic farce is made more intriguing by its unique spatial organization: all three plays cover the same time period and involve the same characters, but each one covers the events in a different part of the house (the dining room, the living room, or the garden). An exit in one play is an entrance in one of the others. The acts of the plays are carefully labeled with the date and time, and the action is exquisitely coordinated so that all three can be assembled (in the viewer's mind) into a single multistage dramatic presentation.

Even though each of the plays makes sense in itself, the actors' frequent exits and entrances arouse our curiosity about what has been going on in the other spaces. For example, the actors enter laughing or express horror at something that has happened in the next room. This is a dramatic structure that cries out for a navigating viewer, but it also suggests the difficulty of telling a story with simultaneous actions. Part of the charm of each play comes from not letting the viewer know things, from showing us effects before we see the causes. At breakfast in the dining room (Play No. 1), everyone is furious with Norman for something he has done the previous night in the living room (Play No. 2). If we had been permitted to follow the characters into the living room, the breakfast scene would lose some of its tension for us. Furthermore, there are times in each of the plays when a character is alone in the space and has to do some bit of stage business to pass the time before the next actor's entrance. A little of that goes a long way. A digital playwright would have to be even more clever than the extremely clever Ayckbourn in order to arrange the dramatic action with such precision that it could sustain suspense over multiple pathways.

A narrative world with such extraordinary spatial depth and temporal continuity is more immersive, more reinforcing of our belief, than a conventional play. However, we will have to develop clear conventions for navigating it. Do we allow movement backward as well as forward in time? Do we allow movement across the same moment of time? Since these will be artistic choices dictated by the story material, there will not be a single answer, but every narrative will have to signal the reader very carefully about what is allowed in order not to raise inappropriate expectations.

The kaleidoscopic stories written by students in my interactive narrative course offer some interesting approaches to this question. In one the reader is placed in a restaurant, and the waiters serve as the means of moving the reader from one group of characters to another. Since no one goes out the door in the course of the story, the fictional world has clear boundaries and does not raise any expectations it cannot meet. In another story, the reader is on a bus that moves forward in screens of text that wipe from right to left across the screen. People get on and off the bus, and the reader can enter their thoughts (including their observations of one another), which form the action of the story. The story offers readers a chance to do something that most bus riders fantasize about: find out who that odd person across the aisle actually is. It enhances its readers' curiosity by not allowing any backward movement. They must follow the bus forward in time, but at each stop they can change their point of view if they wish. In order to find out the whole story, they have to take the trip again, making different choices.[2]

A third student story works like a multidimensional Robert Altman film, focusing on intersecting stories all happening in Manhattan on the same day. A bus driver is having trouble with his eyes, a limo driver is picking up a movie star, a high school couple is going climbing, a young girl is going out with her young aunt. At the points where these stories converge—the bus and the limo are in an accident, the teenagers see the little girl throw up on the subway—the reader can move into the other narrative, much in the way that a movie might use a curious glance from one character to cut into the story of another. Since all the characters are moving around, their stories can be represented on a map of New York, with each scene in the narrative tied to a specific place. The image of the intersecting lines—something like a subway map—tells readers where they are and what they can choose to see.[3]

As the art of digital storytelling develops, authors will gain more skill in fulfilling the possibilities for interesting juxtapositions. Film is an important model for this technique. In the 1920s the Russian film pioneer Lev Kuleshov demonstrated that audiences will take the same footage of an actor's face as signifying appetite, grief, or affection, depending on whether it is juxtaposed with images of a bowl of soup, a dead woman, or a little girl playing with a teddy bear. Using the computer, we can make use of the Kuleshov effect to create juxtapositions that are intentionally open to multiple meaningful interpretations. In a kaleidoscopic story with multiple points of view, any shared event can take on different meanings, depending on whether the same moment is approached in the context of one character's life or another's. The discovery of an infidelity, for instance, might be represented by a neutrally dramatized scene—perhaps a wife reaching for the playback button on an answering machine in the presence of her guilty husband—that would gain in narrative significance when reached from different paths. One sequence might involve the comic recklessness of the husband; another the narcissistic self-absorption of the wife; a third the future devastation of the child, observing this pivotal moment unseen from the doorway. Told in this way, the story would be about the intersecting emotional planes of family experience.

By experiencing such interwoven stories as one unit, we can enhance the kaleidoscopic capacity of our minds, our capacity to imagine life from multiple points of view. One moment in a student story particularly illuminates for me the potential of the medium for capturing delicate dramatic moments with new immediacy. In Rachel Molenaar's "Evening," which takes place in a small apartment, the reader is able to move into the minds of a working mother, her son and daughter, and their pet cat and dog as they go through their nightly routine a few months after the death of the father of the family. At one point they are all relaxing together in front of the TV, parent and children sharing a fragile communion after a tense meal. The mother, increasingly refreshed as the sitcom continues, has a rush of sympathy for her daughter, with whom she had been quarreling earlier in the evening; she silently moves the dozing cat from her own lap to her daughter's. We see this tiny moment of connectedness from three points of view: the mother's, the daughter's, and the cat's. This is an intimate moment that might be presented in linear fiction, as a domestic stream-of-consciousness narrative in the style of Virginia Woolf. It could also be done as a telling dramatic gesture on the stage or on film. But in digital form it takes on a different power. The act of navigating from one consciousness to the other reinforces the separateness of the three fragile creatures and reenacts the gesture of connection. We are in the apartment with them; we see them with the exterior clarity of a film and the interiority of a novel. Such an expressive moment marks the emergence of a new narrative convention, which we might call a panoramic close-up (building on film techniques) or a composite epiphany (building on short-story aesthetics). By rotating our point of view at a single moment of dramatic illumination, we capture both the shared reality and the separate experiences that compose it.

The kaleidoscopic power of the computer allows us to tell stories that more truly reflect our turn-of-the-century sensibility. We no longer believe in a single reality, a single integrating view of the world, or even the reliability of a single angle of perception. Yet we retain the core human desire to fix reality on one canvas, to express all of what we see in an integrated and shapely manner. The solution is the kaleidoscopic canvas that can capture the world as it looks from many perspectives—complex and perhaps ultimately unknowable but still coherent.

Morphing Story Environments

Another kind of narrative experience that takes advantage of the shape-shifting digital medium is one in which interactors are invited to enact or construct their own stories out of a set of formulaic elements. We can think of such an environment as not so much a story as a narrative world that is capable of supporting many possible stories. Brenda Laurel has suggested that virtual reality is not for passive entertainment but for active free-form play and that adults are capable of such play if their imaginations have been "booted up" by an environment with rich narrative possibilities. In chapters 7 and 8 we will look at some of the computational and literary techniques that might allow us to build such environments. For now, let us assume that we have a kind of animated doll's house with characters who are as interesting to adults as the characters on the TV series *NYPD Blue* or *ER* or *One Life to Live*. Suppose we could enter such an environment not merely as players in a MUD but as a god in control of assembling all the characters. Suppose we could set the story of our choice in motion and then observe or participate in it, slipping in and out of whichever roles appealed to us. What would such narrative play be like? Where would it lead to?

Virtual Reality in Haworth Parsonage

It is only the relatively few adults who are gifted storytellers who can spend many waking hours a day in imaginary worlds of their own creation. Perhaps the best-documented example of prolonged imaginative play beyond early childhood is the Brontë family. Left motherless when they were under five and then deprived of their motherly older

sisters, who both died of consumption a few years later, the four sur-
viving Brontë children (Charlotte, Branwell, Emily, and Anne)
amused and sustained themselves by making up stories. They started
with a box of wooden soldiers, which they divided up, named, and
imagined in an African city they called Verdopolis. They played with
the soldiers in the usual way of children, and they made believe that
they were the characters. Soon they started writing a newspaper and
a chronicle of Verdopolis.[4]

Because their older sisters, Maria and Elizabeth, took sick at
school, the Brontë children imagined nightmare schools where children were tortured. When Charlotte was thirteen she wrote down
one such precocious gothic fantasy:

> In the hall of the fountain, behind a statue, is a small door over which
> is drawn a curtain of white silk. This door when opened discovers a
> small apartment, at the farther end of which is a very large iron door
> to a long passage. At the end is a flight of steps leading to a subter-
> ranean dungeon, a wide vault dimly lighted by a lamp which casts a
> death-like melancholy luster over a part of the dungeon, leaving the
> rest in the gloomy darkness of midnight. In the middle is a slab of
> black marble supported by four pillars. At the end of it stands a throne
> of iron. In several parts of the vault are instruments of torture.
>
> At the end of this dungeon are the cells which are appropriated to
> the private and particular use of naughty children. These cells are
> darkly vaulted and so far down in the earth that the loudest shriek
> could not be heard by an inhabitant of the upper world. In these, as
> well as in the dungeon, the most unjust torturing might go on without
> any fear of detection, if it were not that I keep the key of the dungeon
> and Emily keeps the key of the cells and of the huge iron entrance
> which will brave any assault except with lawful instrument.[5]

The Brontë children invented their own *Zork*, a dungeon world to
which they held the key. They were not merely adventurers in this
macabre world, they were the dungeon masters in control of the

instruments of torture and death, masters of the graveyard. The documentation of such a world, the explicitness of its physical dimensions, is reassuring in the same way the solving of a dungeon maze is reassuring. But unlike Zork, Verdopolis was not a single-use toy. It was an open-ended fantasy that changed as the children's emotional needs changed.

When Charlotte went off to school at age fifteen, the game was rearranged to suit the siblings' new obsessions. Emily and Anne, the younger children, rebelliously seceded and invented their own fantasy kingdom of Gondal, ruled by an imperious queen with whom Emily was closely identified. Charlotte and Branwell continued to collaborate on variations of the original story, variations in which Branwell invented a civil war between the main Verdopolis (now Glass-Town) hero, Arthur Wellesley (now Zamorna, King of Angria), and his foil, an ex-pirate and rebel named Percy, Earl of Northangerland, with whom Branwell identified very strongly.

Charlotte, for whom going off to school revived memories of the mistreatment and patient suffering of her sister Maria, invented docile and doomed wives for Zamorna whose names echo her idealized sister's. Marian Hume and Mary Henrietta Percy are beloved daughters and adoring wives who (like Griselda of medieval legend) sweetly submit to their husbands' authority, give up their children without a murmur, and die when Zamorna grows tired of them. Reading about these Mary figures is like watching a child play with a doll so beautiful that she has to be careful not to wrinkle its clothes. They are avatars operated at a distance, repeating a rigidly limited, frightening story.

At this point in the collaborative creative efforts of the Brontë siblings, Branwell was intensely involved with the military strategy of his heroes and their ever-repeating, never-resolved combat, much like a twentieth-century teenager playing videogames, while Charlotte was doing the equivalent of playing with cutout dolls or Barbie dolls, an occupation that has also been transferred into digital form. The difference between the Brontës' play and these less creative activities is that the Brontës had strong narrative imaginations and could assimilate formulaic patterns and appropriate them to their own use. There is nothing original in the Mary doll figures, but they are intensely Charlotte's because she constructed them. The exercise of constructive agency on external, formulaic materials invests the character or toy with the power of a threshold object, the power to create the immersive trance. Charlotte was a young adult with a writer's gift for shaping her make-believe into coherent stories. Her skillful activity is therefore similar to what an unskilled adult might do with a romance novel that is presented not as a paperback but as an electronic construction kit.

Having set a predictable pattern in place, Charlotte began to experiment with variations by inventing foils for her passive heroines. Zamorna's first wife, Marian Hume, has a foil and enemy in Zenobia Ellrington, an accomplished musician and a learned and articulate woman who, as her name indicates, shares with Zamorna an aggressive sexuality.[6] In creating this character, Charlotte raised the question of whether another kind of woman might fit into a world where desire is given free rein. But she kept herself very distant from her creation, treating Zenobia as a clear villain and having Zamorna repeatedly humiliate her.

After trying out these characters and juxtaposing them, Charlotte created a heroine with whom she could identify more closely—Mina Laury, who becomes Zamorna's mistress. Like Zenobia, Mina is characterized by her masculine accomplishments: she can run an estate, counsel political ministers, and travel with the hero in wartime. But she manages to be both assertive and loved. She is overtly rivalrous with Mary and is allowed to shine at Mary's expense, as in this argument between them over who best loves Zamorna:

"I would die for him." [boasts Mina]

"You would not," said [Mary]. "I could not do more and I am confident there is not a woman on earth would do so much for him as I myself!"

"Delicate soft-bred, brittle creature," returned she with kindling eyes, "that is an empty boast. The spirit might carry you far but the body would break down at last . . . it is not for an indulged daughter of aristocracy . . . to talk of serving Zamorna. She may please and entertain him and blossom brightly on his smiles, but when adversity saddens him . . . I warn you, he will call for . . . one . . . who knows the feel of a hard bed and the taste of a dry crust, who has been rudely nurtured and not shielded like a hothouse flower from every blast of chilling wind. . . . With what morbid delicacy would you shrink from scenes that I have looked upon unmoved."[7]

Although she was offering to die for Zamorna, Mina's speech is in fact a declaration of survival. She was saying that it is better to be coarse and passionate and alive (as Charlotte thought of herself) than to be delicate and docile and dead (as she thought of her sister Maria). Mina was a clear step in the direction of the defiant and sturdy Jane Eyre, the woman who ministers to the rough and surly Rochester and copes with fire, stab wounds, and secret midnight errands. As Charlotte matured, she transformed her doll-like heroines into the shape of women she could grow into. Her stories were her way of rehearsing for the emotional hardiness she would need in her adult life.

The regressive, violent, overheated emotional universe of the young Brontës is very like the narrative world of many electronic games. *Myst*, for example, is also the result of a long sibling collaboration (between Rand Miller and Robyn Miller) and has the same Napoleonic echoes, the same bloodthirsty playacting quality, and the same emphasis on betrayal and power struggles as the Brontë juvenilia, though it is set in magical worlds with sentient monkeys and a science fiction veneer. One element that renders both the Brontë juvenilia and the Miller brothers' fantasies so claustrophobic is the undisguised nature of their wish fulfillment. *Myst*'s primitive vision of evildoing and rich rewards is appropriate to the current early stage of electronic fiction development. In fact, part of the reason the "winning" ending of *Myst* is so dull is that it does not partake of this gothic, violent undercurrent that runs through the rest of the game. In order for electronic narrative to reach a higher level of expressiveness, the medium as a whole must make the shift that Charlotte made, that is, away from adolescent rehearsal fantasies and toward the expression of more realistic desires. In Charlotte's case we have a record of how she made the shift, one that suggests how constructive interactors might use such overheated participatory narratives.

Charlotte was the only one of the four Brontë children to end her childhood fantasy life when she reached adulthood. Branwell continued to sign himself *Northangerland* to the end of his life. Emily and Anne played at their fantasy kingdom until their late twenties (when Charlotte disrupted it so that they could all write for publication). But Charlotte abandoned Zamorna and his world at age twenty-three. She was able to do this because she had taken her Angria fantasy as far as it could go in exploring her feelings of sexual longing and guilt. In one of her last Angrian stories, *Caroline Vernon,* the deepest level of wish fulfillment within the fantasy comes to the surface in very explicit terms, and the protagonist is as closely identified as possible with the author. Not only is *Caroline* a variant of *Charlotte,* but Caroline, like her author, has a "delicious" obsessive daydream of a fantasy lover, "a hero, yet nameless and formless, a mystic being, a dread shadow . . . [who] haunted her day and night when she had nothing else useful to occupy her head or her hands." Charlotte is here confronting the absurdity of her own infatuation with the imaginary Zamorna and is allowing herself to look at the boundary between her immersive world and her ordinary life. She is getting ready to leave her make-believe kingdom.

In *Caroline Vernon* the underlying sexual content of the Zamorna/Percy/Mary plot becomes clear: the wish to supplant the mother/sister in the regard of the father/lover. Caroline is the illegitimate child of Percy and is rivalrous with her legitimate sister Mary for their father's love, as well as for the romantic attention of Zamorna, who is Caroline's "guardian" and Mary's husband. Caroline's mother, the actress Louisa Vernon, is romantically involved with both Percy and Zamorna. The sexual rivalry of mother and daughter and the mother's antagonism toward the daughter's sexuality—both classical oedipal configurations—are presented in *Caroline Vernon* in much more explicit terms than in any of the Brontës' earlier stories. They are magically resolved in favor of the daughter, as in this scene in which Louisa threatens to tell Percy that Caroline has been flirting with Zamorna:

> "I'll tell you all!" almost screamed her ladyship. "I'll lay bare the whole vile scheme! your father shall know you, Miss, what you are, and he is. I never mentioned the subject before, but I've noticed, and I've laid it all up and nobody shall hinder me from proclaiming your baseness aloud."
>
> "Good heaven! this won't do," said Caroline, blushing red as fire. "Be silent, Mother. I hardly know what you mean, but you seem to be possessed. Not another word now. Go to bed—do. Come, I'll help you to your room."
>
> "Don't fawn; don't coax," cried the infuriated little woman. "It's too late. I've made up my mind. Percy, your daughter is a bold, impudent minx; as young as she is, she's a————"
>
> She could not finish the sentence. Caroline fairly capsized her mother, and took her into her arms and carried her out of the room. She was heard in the passage calling Elise and firmly ordering her to undress her lady and put her to bed. She locked the door of her bedroom and then she came down stairs with the key in her hand.[8]

The scene has the clarity and absurdity of a dream, a simple fulfillment of an infantile wish to change places with mother, to pick up the "little woman" and put her to bed. Now it is the hostile and punishing mother who is locked up in the dungeon, and the daughter is free to do what she pleases. The story ends with Caroline seduced by Zamorna, and civil war opening all over again between Zamorna and Percy. The story cannot go any further. It has come clear. At this point Charlotte bids farewell to Angria and turns her attention to writing realistic fiction.[9]

Charlotte's departure from Angria is one model for achieving closure in a shape-shifting story world. The experience of the underlying
fantasy coming to the surface is not merely an exhaustion of narrative
possibilities; it is more like the solution to a constructivist puzzle. Projection of highly personal (but universally felt) emotional content
onto the figures of the formulaic story moves the content into a field
where it is safe to think about it. It is putting your most dangerous
fantasies in a dungeon to which you hold the keys. Because the fantasy has been externalized, it can be manipulated. If the external
structure is too rigid and too literal (like Branwell's endless military
campaigns or the endless combat of the video arcade), the externalized fantasy can serve as a safety valve but the plot does not progress
and the experience does not reach closure. But if the imaginative environment is more plastic and more ambiguously evocative, the fantasy can progress. We can sustain our engagement in such a
constructivist world, bring our deepest emotional conundrums into
it, and then play them out in multiple ways until they come clear. The
experience of closure here may not be in the beauty of the particular
story but in the completeness of engagement with the whole range of
story possibilities.

D. W. Winnicott described a similar process of imaginative "saturation" in children's play. Children play with a certain toy or play out a
certain imaginative experience until it has absorbed all the emotional
ambivalence they feel about the subject, and then they are ready to
transfer their feelings to the world at large, to what Winnicott called
the "whole cultural field."[10] As a society we use television series in
much the same way, asking them to present us with situations that
are particularly frightening or appealing (crime, emergency rooms,
family life, women in the workplace, sexually active single people)
and that we have not yet assimilated into our national consciousness.
The programs assemble formulaic characters and situations that express our anxieties and desires and then offer variations on the patterns. When every variation of the situation has been played out, as
in the final season of a long-running series, the underlying fantasy

comes to the surface. Thus, the Miami cop suffers amnesia while doing undercover work and becomes a criminal, the crusty newsroom boss imagines he is actually sleeping with the perky producer, the mismatched couple who have resisted one another finally get together. These episodes are often embarrassing in the same way that *Caroline Vernon* is embarrassing. Robbed of the elaboration of sublimation, the fantasy is too bald and unrealistic, like the child carrying the mother up to bed. The suppressed fantasy has a tremendous emotional charge, but once its energy has saturated the story pattern, it loses its tension. We can look at it directly, with less anxiety, but we also find it less compelling. Nevertheless, such formulaic stories can also be very insightful and well written before they run out of steam. In fact, their exhaustion can be an indication of the fact that we have assimilated their patterns into our general understanding of the world, that there is no longer a story for us in the core situation, whether it be people dying in a hospital room full of strangers or aggressive women in the work world or frighteningly dysfunctional child rearing. The shifting patterns of the formulaic story have expressed our ambivalent feelings, absorbed our excitement, and made the threatening or alluring situation into something familiar.

Enactment as a Transformational Experience

As Scheherazade and Jesus both knew, storytelling can be a powerful agent of personal transformation. The right stories can open our hearts and change who we are. Digital narratives add another powerful element to this potential by offering us the opportunity to enact stories rather than to merely witness them.

Enacted events have a transformative power that exceeds both narrated and conventionally dramatized events because we assimilate them as personal experiences. The emotional impact of enactment within an immersive environment is so strong that virtual reality installations have been found to be effective for psychotherapy. Psychologists in several research centers are treating phobic patients by exposing them to virtual environments that simulate the situations that trigger their anxiety attacks. The desensitizing process is in essence a participation in a fictional world. Researchers in California and Atlanta have relieved patients' long-standing fear of heights by having them "walk" over virtual bridges and ride in virtual elevators. Patients initially respond to the virtual environments with terror, just as they would to the real-world experience. The therapist then accompanies them through the experience, helping them practice self-calming behaviors. Essentially, the patients are practicing coping behaviors in the virtual environment; they are like actors at a dress rehearsal. The inner changes brought on by such experiential learning then allow them to apply the same behaviors to the real world. Patients who can ride a virtual glass elevator in a virtual hotel lobby can then go to dinner on the seventy-second floor of the Peachtree Plaza in Atlanta, and patients who cross a virtual Golden Gate Bridge can then cross the real one.[11]

This virtual reality therapy falls between two other therapeutic techniques: actually accompanying the patient in the real-life frightening situation and talking the patient through an imaginary experience under hypnosis. The virtual world is more external than the hypnotic experience but artificial enough to make it possible for patients to approach it at a much earlier stage than they could if facing the actual situation. It is a threshold environment. Those patients who do not find VR therapy helpful are those who complain that it is either too real or not real enough. The key to the success of the treatment, then, is the establishment of the world as a fictional space.

These results echo the processes observed in some MUD participants who use their imaginary personas to practice social skills they are trying to cultivate in the "real" (i.e., nonelectronic) world. For instance, one woman recovered her sexual confidence after an amputation by enacting the part of a similarly handicapped character on a MUD.[12] As in the case of the phobic patients, the virtual experience

worked because it was enough like the real one to raise the same anxieties but safe enough to allow for imaginative rehearsal.

The transformational power of enacted narratives holds both
promise and danger for the future. On the one hand, it may make
digital environments as important as television currently is for the
presentation of problem plays, stories about social injustice or intoler-
ance that are meant to broaden the audience's sympathies. Electronic narratives are already being used to teach such skills as
language learning, military medicine, and corporate decision making.
They could also be used to teach ways of being in the world, to teach,
for example, how to resolve conflicts, how to be a successful job ap-
plicant, how to be a nurturing parent, how to be a nonabusive spouse
or parent. If these issues are embedded in interactive narrative that is
fictionalized just enough to be compelling but not threatening, such
narratives might be as effective in changing behavior as an acropho-
bic's walk across a virtual bridge.

On the other hand, computer enactment may also reinforce violent or antisocial behaviors. Already a college student in the Midwest
has been disciplined for publishing on the Internet a rape fantasy in
which he names an actual fellow student. We may be moving toward
a situation like that depicted in the *Star Trek* episode, aptly named
"Hollow Pursuits," in which a withdrawn and awkward crew member
becomes addicted to holodeck programs that allow him to outfight or
seduce the people he is intimidated by in his actual life.[13] Just as psychologists are considering scanning in images of their patient's actual
family members for VR therapy, there is no reason why people could
not scan their boss's image into a customized version of *Doom* and
blast away. Would this exercise make it more or less likely that they
would actually shoot their boss?

Whatever the answer, it is clear that literal wish-fulfillment fantasies would not help a person cope with the actual situation. The
difference between immersive environments that are escapist and
those that are progressive may lie in the difference between the repetitive stories of Branwell Brontë and the progressive stories of Char-

lotte Brontë. The more fully constructivist the story environment, the more opportunities it will offer to move beyond the enactment of destructive patterns. The goal of mature fictional environments should not be to exclude antisocial material but to include it in a form in which it can be engaged, remodeled, and worked through. Therefore, an environment in which we can only kill dragons, no matter how many different ways we can transform their appearance, is less desirable than one in which we can also domesticate them, worship them, ally with them against other monsters, or perhaps even take them for a ride in a multistory atrium elevator.

Refused Closure

In a shape-shifting world, stories often do not come to a clear end point. Electronic narrative teases us, holding back its gifts. The labyrinth is tricky, full of dead ends, uncertainties, questions that do not resolve. Adventure games demand hundreds of hours of play, of mostly frustrating trial and error, to discover the way forward. Sometimes their secrets must be discovered outside the game, from magazines or by trading information with fellow players or perhaps by finding one's way over the Internet to the right Web site or newsgroup. In the solutionless rhizome or the solvable maze, we are confronted with a world that lures us in with the promise of treasures but that is chiefly designed to resist our efforts.

Perhaps this is a virtue. To be always in search of secret information, in pursuit of refused reward, can be emotionally riveting. Because we are aware that this is a created world, we can experience its resistance to our efforts as a dramatic contest with the programmer or writer over a gift that is purposely withheld. We may experience this withholding presence as a demanding parent, a challenging teacher, a coy lover, or a secretive boss, and be all the more engaged by the contest. Or we may experience it as a sustained arousal, a prolonged lovemaking with the climax always a little out of reach. Because of its ability to both offer and withhold, the computer is a seductive medium in which much of the pleasure lies in the sustained engagement, the refusal of climax.

The question of confused extent and refused closure is explicitly posed by Michael Joyce in his hypertext novel *Afternoon,* which has no overview of contents and no clearly marked ending. Instead, Joyce tells his readers to decide for themselves when the story is over. In a lexia entitled "work in progress," he states plainly: "Closure is, in any fiction, a suspect quality, although here it is made manifest. When the story no longer progresses, or when it cycles, or when you tire of the paths, the experience of reading it ends." This is closure as exhaustion, not as completion. Not satisfied with this formulation, others have described closure in *Afternoon* in terms akin to the solution to a maze. For example, it is said to be achieved when "having assigned particular sections to particular sequences or reading paths, many, though not all, of which one can retrace at will, one reaches a point at which one's initial cognitive dissonance or puzzlement disappears, and one seems satisfied. One has reached—or created—closure!"[14]

In other words, electronic closure occurs when a work's structure, though not its plot, is understood. This closure involves a cognitive activity at one remove from the usual pleasures of hearing a story. The story itself has not resolved. It is not judged as consistent or satisfying. Instead, the map of the story inside the head of the reader has become clear. Such a map does not necessarily feel inevitable or appropriate, the way the solution to a puzzle does. It may not be beautiful or shapely in any way. There is no emotional release or perception of fittingness, just a sense of going from the unknown to the known. This is very different from and far less pleasurable than our more traditional expectations of closure, as arising from the plot of the story and marking the end point of an action.[15]

Of course, closure can be feared as well as desired. This is so not only because of disturbing story content, as in *Afternoon,* but also because it is painful to break the immersive trance, to come up from an enveloping medium into the chilly air of reality. In encyclopedic narratives in particular—a three-decker novel, a television series, a months-long puzzle game—the ending can be painful to the creators and to the audience. Dickens and his audience both cried when the last number of one of his two-year serials was finished. When the television series *Cheers* ended, it provoked an orgy of public nostalgia, as if the actual neighborhood bar of millions of people were closing.

The refusal of closure is always, at some level, a refusal to face mortality. Our fixation on electronic games and stories is in part an enactment of this denial of death. They offer us the chance to erase memory, to start over, to replay an event and try for a different resolution. In this respect, electronic media have the advantage of enacting a deeply comic vision of life, a vision of retrievable mistakes and open options. The never-ending, ever-morphing cyberspace narrative is a place to revel in a sense of endless transformations, but in order for electronic narrative to mature, it must be able to encompass tragedy as well.

Tragedy in Electronic Narrative

How do we express the irreparable losses of life with appropriate solemnity within a shape-shifting world? How can we have catharsis in a medium that resists closure? Since no hypertext story or simulation narrative in this early stage of genre development offers a satisfyingly tragic story, we can only imagine them. In my use of the word *tragic*, I am relying on Aristotle's definition, as it is most commonly and somewhat loosely understood. I mean a story of a single worthy individual's fall from a worthy life to a desperate ending through some choice or flaw of his own, a story that focuses on this irretrievable loss, arousing our feelings of pity and terror and leaving us at the end in a state of purged emotion and heightened understanding.

Let us consider the representation in electronic form of the tragic event of a young man's suicide. I want to propose three possible suicide stories, each about the same fictional character, whom I'll call Rob. Each story exploits the multiform nature of electronic narrative while still giving expression to the tragic heart of the story.

The Mind as Tragic Labyrinth

Suppose someone were to write an electronic portrait of Rob's mind on the night of his suicide, like a stream-of-consciousness novel but in the form of an animated web. Each lexia, or unit of the story, would capture one haunting thought and would link to one or more others. Rob's mind would run obsessively, from recent events (a bad performance review, a lost promotion) to past disasters (his father's praising his younger brother for beating him at chess, his ex-girlfriend Linda's decision to move out). Some thought patterns would lead to grandiose hopes for money and fame or to ambitions of perfection. All these chains of thought would eventually converge on the idea of killing himself. Thoughts of going for help could be represented by false links; you could click on them, but the screen would not change (they would lead nowhere). Rob might experience short flashes of remembered moments in which life was good, but these would be followed by dead-end beliefs that such moments of happiness have been lost forever or have all been invalidated by later failures or betrayals. Perhaps the navigating reader would feel impelled to return to a good memory or to trace it more deeply but would find those associations closed off, blocked by unpleasant thoughts, or too difficult to hold on to. Perhaps the accounts of good memories would fade quickly from the screen, or perhaps other, destructive, thoughts would intrude involuntarily, as represented by images or scenes that would arise by themselves without any action from the reader. All these phenomena—the obsessive rumination on failure, the doomed grandiosity, the destructive thoughts coming unbidden—would provide the reader with the opportunity to enter the consciousness of the desperate man.

By charting the thoughts of such a complex person, the writer could take as her subject matter the actual process of rumination— the repeated return to associational paths that lead to closed loops of thinking or the poignancy of a single moment of experience, a single act of perception that becomes lodged in the mind, like a roadblock on the path to hopefulness. A labyrinthine hypertext might be the ideal medium for capturing the interior monologue as a sort of snapshot of the mind itself. It could potentially offer the equivalent of Hamlet's "to be or not to be" soliloquy, not as a translation of it but as a similarly affecting universal portrait of paralysis and self-awareness.

How would the interior monologue reach climax and closure? The paths of the mind could change as these simulated last hours of Rob's life progressed, so that thoughts of suicide would arise more quickly each time regardless of which paths the reader followed. Perhaps the computer would allow the process to continue only until a set number of lexias of suicidal ideation have been selected by the reader or until a certain frequency of access occurs. Suddenly, the screen darkens. The suicide occurs in "real" time. The reader would have both enacted and witnessed the decision and would feel the sense of understanding, inevitability, and sorrow that we call catharsis.

The Web of Mourning

Another way of representing the same story might be as a three-dimensional presentation of the wake, with Rob's devastated parents, his coworkers, his ex-girlfriends all trying to understand how such a tragedy could have happened. Each would try to explain the suicide, and by moving around the room from one mourner to another, we could follow their memories and see events from the past from their perspectives as people engaged in the inevitable but futile attempt to fix a cause for an irrational act. We would be drawn to see the whole story, since no one version would account for the event but each would add to our understanding.

In navigating this kaleidoscopic story, we would realize that each individual version includes elements that do not quite fit or that point us away from the teller's version to other explanations. We would experience the loss in all of its resonance and have a sense of all the worlds of caring and trust that are torn apart by a violent death. Going through such a story would be an enactment of mourning, because the interactor's search for an explanation would parallel the efforts of the people at the wake. Perhaps we would be given only a set number of scenes to witness at any single viewing. Then we would see the burial, with the program perhaps taking us through that final scene from the perspective of whichever mourner we had spent the most time with. Each separate viewing would provide its own experience of catharsis, but no single one would feel complete. Only after viewing all the stories, after repeating the mourning process from each of the several viewpoints, would we feel a larger catharsis: not an acceptance of Rob's death, not an understanding of a single consistent composite explanation, but a pervasive sense of an interrelated community with multiple truths. After tracing the multiple contexts for a single act of suicide, we would be left with a tragic vision of the many Robs who had been lost.

Simulation and Destiny

These stories take the events of the story as fixed. We can trace them in multiple ways, but we cannot act within them or change the plot. Could a digital narrative offer a higher degree of agency while still preserving the sense of tragic inevitability? Can we have an interactive story that still retains what Umberto Eco calls its sense of destiny?[16] One test of the limits of electronic narrative to provide agency and destiny would be to consider whether we could frame the story of Rob's suicide in the form of a simulation.

In the simulation treatment of Rob's suicide, the interactor would be put in the position of a god over Rob's social world but a god with a limited power of intervention. The presentation of the world would make clear the limits of our powers, would specify what elements in Rob's world the author is taking to be immutable. These immutable elements would be an important literary choice, for they would characterize an implicit interpretation of the "why" of Rob's death. Would it be his neurochemistry? His family history? A particular historical moment or political situation? A single tragic love affair? Varying the other factors would allow Rob's life to play out in Borges' "pullulating" time, with the fascinating diversity of possibilities that any life is open to. As deity interactor, we might guide Rob to a different job with a manager who would be more appreciative of his particular skills. Or we might send an old girlfriend to a restaurant just as Rob arrives there, to see if they could restart their relationship. Maybe we would arrange for a friend of Rob's mother to encourage her to confront her husband about his constant put-downs of their son. One by one we could change all of the contingent conditions that seem to contribute to Rob's self-destruction. Seeing how his life would play out in different ways would add to our sense of his vulnerability, of what little influence external factors have. We would also see how changing Rob's life would dramatically alter the lives of the people around him. Perhaps he would have been more miserable rather than more hopeful if his old love affair had resumed. In that alternate world we might see his girlfriend blaming herself for his death, not knowing that it was inevitable. Or we might reach some ambiguous ending in which it is possible to believe that Rob will be revived in time to survive or even that he might not go through with the suicide attempt. The story would still have a sense of completion despite this variation because it would focus not on a single continuous action (as Aristotle described Greek plays) but on the whole system of subtle interrelationships that give meaning to one person's action. This narrative system would reveal itself through its range of variation as well as through what remains constant. The tragedy of the situation would arise from a demonstration of the ways in which people unwittingly play into destructive patterns, sometimes from the best of intentions.

Of course, the success of any of these possible tales would depend on the writer's skill and on the particulars of storytelling that I am just crudely sketching. But there is no reason why a story like this could not be as expressive as tragic stories in any other medium. What is more, a digital narrative could capture something we have not been able to fix as clearly in linear formats: not just a tragic hero or a tragic choice but a tragic process.

The Multipositional View

Whether we are talking about a simulation story, a rhizome hypertext, a navigable movie, or an electronic construction kit for never-ending stories, we cannot bring to a transformative, shape-shifting medium the same expectations of static shapeliness and finality that belong to linear media. But that does not mean that we will forgo a sense of completeness and emotional release. Instead, we will learn to appreciate the different kinds of closure a kaleidoscopic medium can offer.

Simulation exercises offer a good model for kaleidoscopic closure. When a live-action simulation ends, the participants hold a wrap-up session in which the god of the machine—the game master or controller of the simulation—describes what has happened and solicits experiences from all the players, who now have the opportunity to see how their individual parts fit into the overall story and to understand the many processes that make up the microworld of the simulation. For instance, in a training situation that models a foreign affairs crisis, the participants would learn what their opponents' motives were, what other options they had been considering, and what information was known to whom at various stages of the game. In a narrative game, much of the drama of the story is not apparent until the wrap-up, when there are important plot revelations, such as who actually committed which murders, which players were long-lost secret brothers, and where the magic sword really was at 11:00 P.M. on Saturday night. At the end of the game players are able to see the whole action of the story, including their own part in it, not from the stage but from the perspective of a spectator at the top of the arena.

It is satisfying to switch positions in this way, to act in a patterned event and then later view the general pattern, like a synchronized dancer in one of the old Ziegfeld dance movies watching footage of an overhead shot of her number. But a computer simulation offers a new extension of this pleasure. On the computer we can reenter the story and experience more than one run of the same simulation. We can play all the parts, exhaust all the possible outcomes. We can construct a composite view of the narrative world that does not resolve into any single story but instead composes itself into a coherent system of interrelated actions. Because we increasingly see the world and even our own identities as such complex, centerless, open-ended systems, we need a story environment that allows us to make sense of them by enticing us into exploring a dense narrative world from every possible perspective.

One of the results of such an exploration will be a more immediate appreciation of process. Whereas novels allow us to explore character and drama allows us to explore action, simulation narrative can allow us to explore process. Because the computer is a procedural medium, it does not just describe or observe behavioral patterns, the way printed text or moving photography does; it embodies and executes them. And as a participatory medium, it allows us to collaborate in the performance. Using the computer, we can enact, modify, control, and understand processes as we never could before. We can also appreciate them aesthetically for the first time, savor the complex patterns of processes just as we savor patterns of color and shape. We do not yet have story systems that exploit this potential by describing a complex world in the procedural terms, but we are moving steadily in that direction.

The three aesthetic principles described in this section—immersion, agency, and transformation—are not so much current pleasures as they are pleasures we are anticipating as our desires are aroused by the emergence of the new medium. These pleasures are in some ways continuous with the pleasures of traditional media and in some ways unique. Certainly the combination of pleasures, like the combination of properties of the digital medium itself, is completely novel. To satisfy our desire for this new combination of pleasures, we will have to invent techniques of authorship that are similarly eclectic. The next section deals with the merger of literary and computational techniques of composition. It explores how we might go about writing for a process-centered medium and what stories and characters such procedural authorship will bring us.

PART III

Procedural Authorship

ALAS, POOR YORICK! I KNEW HIM,
HORATIO: A FELLOW OF INFINITE JEST, OF
MOST EXCELLENT FANCY: HE HAT BORNE
ME ON HIS BACK A THOUSAND TIMES;
AND NOW, HOW ABHORRED IN MY
IMAGINATIOIOIOOOIOIOIIOIOIIIIIOIOIOIO
OIOIOIOOIOIOIIOIOOOIOIOIOIOIOIOIOIOO
OOOIOIOIIIIOIOIOIOIOIOIOOIOIOOIOIO
OIOIOIOIOIIIOIOIOIOOOIOIIIIIOIOIOOIII
OOIIIOOIOIOIOOIOIOOOIOIOIOOIOIOI
OIOOOOIOIIIIIOIOIOIOIIOOOOOIIIIIIOIOI
OIOIOIIIOIOIOIOIOIOIOIOIOIIIIIOIOOOO
OIOIOIOIOIOOIIIIOOIOOIOOOOOIOIIOOO
OIOIOIIIIOIOIOIOIOIOIOOIIIIIOIOIOIOI
OOOIIOIIIOIIIIIOIOIOIOIOIOIOIOIOOOI
OIOIOIOIOIOIIOIOOOIOIOIOIOIOIOIOOO
OOIOIOIIIIOIOIOIOIOIOIOIOIOIOOIOIOOI
OIOIOIOIOIIIOIOIOIOOOIOIIIIIOIOIOOIIIOO
IIIIOOIOIOIOOOIOIOOOIOIOIOOIOIOIOI
OOOOIOIIIIIOIOIOIOIOIIOOOOOIIIIIIOIOIOI
OIOIIIOIOIOIOIOIOIOIOIOIOIIIIIOIOOOOOI
OIOIOIOIOOIIIIOOIOOIOOOOOIOIIOOOIO
IOOOIOIIIIOIOIIOIIIOOIIIIOOIOIOIOOOIO
IOOOIOIOIOIOIOIOIOIOOOOIOIIIIIOIOIOI
OIIOOOOOIIIIIOIOIOIOIOIIIOIOIOIOIOIOI
OIOIOIIIIIOIOOOOOIOIOIOIOIOIOOIIIIIOOI
OOIOOOOIOIIOOOIOIOOOIOIIIIIOIOIOIO
IOIOOIIIIOIOIOIOIOOOIOIOIOIOIOIOIOO
IOIOOIOIOOIOIOIOIOIIOIOIOIOOOIOIIIIII
OIOIOOIIIOOIIIOOIOIOIOIOOOIOIOOOIOI
OIOOIOIOIOIOOOOIOIIIIIOIOIOIIOOIIIIO
OIOIOIOOOIOIOOOIOIOIOOIOIOIOIOIOOO
OIOIIIIOIOIOIOIOIOOOOOIIIIIOIOIOIOIOIII
OIOIOIOIOIOIOIOIOIIIIIOIOOOOOIOIOIOI
OIOOIIIIIOOIOOIOOOOOIOIIOOOIOIOOOI
OIIIIIOIOIOIOIOIOOIIIIIOIOIOIOIOOOIOIO
IIOIOIIIIIOIOIOIOIOOIOIOIOIOOOIOIOIOOI
OIOIIOIOOOIOIOIOIOIOIOIOOOOOIOIOII

ALAS, POOR YORICK! I KNEW HIM, HORATIO; A FELLOW OF INFINITE JEST, OF MOST EXCELLENT FANCIES; HE HAT BORNE ME ON HIS BACK A THOUSAND TIMES; AND NOW, HOW ABHORRED IN MY IMAGINATION IT IS!

Chapter 7
The Cyberbard and the Multiform Plot

A plot is . . . a narrative of events, the emphasis falling on causality. "The king died and then the queen died" is a story. "The king died and then the queen died of grief" is a plot.

—E. M. Forster, *Aspects of the Novel*

What will it take for authors to create rich and satisfying stories that exploit the characteristic properties of digital environments and deliver the aesthetic pleasures the new medium seems to promise us? We would have to find some way to allow them to write procedurally; to anticipate all the twists of the kaleidoscope, all the actions of the interactor; and to specify not just the events of the plot but also the rules by which those events would occur. Writers would need a concrete way to structure a coherent story not as a single sequence of events but as a multiform plot open to the collaborative participation of the interactor. At first this kaleidoscopic composition seems like a violent break with tradition, but when we look at how stories have historically developed, we find techniques

185

of pattern and variation that seem very suggestive for computer-based narrative.

From the nineteenth century on, there has been considerable interest in the striking similarities found between stories from vastly different cultures. Carl Jung hypothesized that these similarities offer proof of a collective unconscious, a set of archetypal tales (the journey, the quest, the rebirth) and archetypal figures (the hero, the trickster, the earth mother) that together define what it is to be human. Some have argued that all of the world's great wisdom stories express the same religious and psychological truths and therefore are just variant versions of a single tale. Joseph Campbell, one of the most passionate and eloquent proponents of the unity of human narrative traditions, saw in stories as diverse as those of Prometheus and Buddha a single "monomyth" about a "hero with a thousand faces" who sets off from the common world, encounters fantastic dangers, and returns to bestow gifts on his society.[1] Although such totalizing views of human culture have fallen out of fashion, the enduring appeal of mythic patterns is indisputable. We have only to look at George Lucas's ubiquitous Star Wars series (which was directly inspired by Campbell's research) to see that age-old story epic formulas remain compelling even in our postmodern and antiheroic era.

But it is not only folktales and adventure stories that have formulaic patterns. Many narrative theorists and writers have insisted that there are a limited number of plots in the world, corresponding to the basic patterns of desire, fulfillment, and loss in human life. Rudyard Kipling counted sixty-nine basic plots, and Borges thought that there were less than a dozen. Ronald B. Tobias, in one of the more competent of the many guidebooks for writers, suggests there are twenty "master plots" in all of literature. Here is his list:[2]

- Quest
- Adventure
- Pursuit
- Rescue

- Escape
- Revenge
- The Riddle
- Rivalry
- Underdog
- Temptation
- Metamorphosis
- Transformation
- Maturation
- Love
- Forbidden Love
- Sacrifice
- Discovery
- Wretched Excess
- Ascension
- Descension

One would be hard put to name any story that did not belong, at least in part, to one of these categories, whether it is *The Incredible Hulk* (metamorphosis), *King Lear* (descension), or *Seinfeld* (refused maturation). The patterns are constant because human experience is constant, and though cultural differences may inflect these patterns differently from one place to another and one historical period to another, the basic events out of which we tell stories are the same for all of us.

The formulaic nature of storytelling makes it particularly appropriate for the computer, which is made for modeling and reproducing patterns of all sorts. But no one would want to hear a story that was a mere mechanical shuffling of patterns. How do we tell the computer which to use and how to use them? How can the author retain control over the story yet still offer interactors the freedom of action, the sense of agency, that makes electronic engagement so pleasurable? To answer these questions we have to look back at an earlier storytelling technology, the community of oral bards.

The Oral Bard as a Storytelling System

We now know that densely plotted, encyclopedic works like the *Iliad* and the *Odyssey* were produced not by a single creative genius but by the collective effort of an oral storytelling culture that employed a highly formulaic narrative system. From the Renaissance through the beginning of the twentieth century, Homer was considered to be a great *writer*, rather than a preliterate singer, and his epics were taken as the height of Western literature. It was therefore disconcerting when the Harvard classicist Milman Parry and his student, Alfred Lord, documented the similarities between the Homeric poems and those they heard performed by oral bards still active in Yugoslavia earlier in this century. Lord's book, *The Singer of Tales*, published in 1960, describes the actual composition and performance process of the bards and argues from internal evidence that Homer's poems are the result of similar methods.

Oral story composition, as Lord describes it, relies on what we in a literate era devalue as repetition, redundancy, and cliché, devices for patterning language into units that make it easier for bards to memorize and recall. The stories are composed anew for each recitation and are therefore multiform, with no single canonical version. Every performance of a story varies from all others, reflecting the interests of the audience and the dramatic interpretation of the storyteller. Lord's research gives us a detailed picture of these multiform stories and how they achieved a coherency of plot across many varied tellings.[3]

The bardic tradition is a set of formulas within formulas, starting at the level of the phrase and moving through the organization of the story as a whole. Singers of tales had a repertoire of formulaic ways to describe common people, things, and events, descriptions that could be rearranged and plugged into a template of the chanted line in a way that made for pleasurable variation within an overall pattern of regular rhythms and sounds. Major characters were associated with familiar epithets that helped recall them to the audience and fill out the poetic line; for instance, Homer could refer to Zeus as "the coun-

selor" or "son of Chronos" or "the cloud-gatherer." Lord discovered
that a hero might have one typical epithet when invoked in the beginning of a line and another one more commonly used after the
caesura in the middle of the line. The appellation would therefore
have less to do with how the hero was behaving at a particular moment in the narrative than with where his name fell in the rhythm.

Lord refers to the bard's stock of variant phrases as a "substitution
system." This is similar to the Mad Libs parlor game, where a paragraph is given with words missing and players are asked to contribute
words based on syntactic or category descriptions (a noun, a body
part, a furry animal, etc.) with hilarious results. Early attempts at
computer-based literature tried to use similar methods of simple substitution, with equally incongruous results. For instance, here are two
of the millions of possible stories based on a language substitution system devised by the French experimental writer Raymond Queneau:

> *A Story As You Like It, version #1:*
> Do you wish to hear the story of the three big skinny beanpoles?
> The three big beanpoles were watching them.

> *A Story As You Like It, version #2:*
> Do you wish to hear the story of the three middling mediocre
> bushes?
> The three middling mediocre bushes were watching them.
> Seeing themselves voyeurized in this fashion, the three alert peas,
> who were very modest, fled.[4]

Although such compositions are provocative as artifacts that play
with the theoretical concepts of literature, machines, and originality,
no one would read such stories for the sheer pleasure of it. But Web
surfers can now visit many entertaining sites where they can generate
endless parodies of corporate home pages, candidate's speeches, and
even love letters by using substitution programs stocked with the appropriate buzzwords. These pages provide the same pleasure as

ELIZA (see chapter 3); they use the rote utterances of the machine to expose the meaninglessness of such formulaic writing.

But even if a verbal substitution system cannot by itself produce satisfying and coherent digital narratives, it is a useful model for establishing the "primitives" or basic building blocks of a story construction system. In computer programming systems the "primitives" are the smallest components (such as simple arithmetical calculations) upon which the larger operations (such as complex calculus functions) are built. In an interactive narrative the key primitives are the actions of the interactors themselves, as structured by the author. Currently, the most complex sets of primitives are the icon palettes in puzzle games, which contain items like a magnifying glass, a set of objects picked up on the way (tools, treasure, food), communications devices, a way to select who should be speaking or acting, and perhaps an icon for a hand or a foot to allow the interactor to pick things up or move around. Such a palette constitutes an iconographic substitution system, since the interactor can substitute one tool for another, or one item for another as the object of the tool, in order to build up more complex possibilities of action, in the same way that the epic singer could substitute phrases in making up his poetic line.

In order for the medium to mature, storytellers will have to develop more expressive primitives, simple actions that will allow ever-subtler input by the interactor. For instance, game designers have already progressed from an interface that requires the user to type "go north" or even "n" to one that allows users to point and click in three-dimensional space, which changes the interaction from a command structure to a dramatic gesture. The easier these primitives are to learn and the less they call attention to the computer itself—that is, the more transparent they are—the deeper our immersion and the stronger our sense of dramatic enactment. Today's interaction conventions are equivalent to the invention of a few useful epithets for the gods and heroes, basic tools that every storyteller needs but not enough to get you very far with a particular tale.

One of the chief stumbling blocks to mature digital storytelling is the difficulty of establishing expressive conventions for the interactor's use of language. If we give the interactor complete freedom to improvise, we lose control of the plot. But if we ask the interactor to pick from a menu of things to say, we limit agency and remind them of the fourth wall. Some CD-ROM stories give the interactor the task of deciding the mood or tone of a spoken response rather than picking a statement from a list of possible things to say. This is a more promising route because it seems less mechanical, although the mood selector is often a menu or slider bar that is outside the frame of the story. A more immersive set of primitives might be applied to character gesture. Gestures like placing a hand on the shoulder of another character, making a fist, raising both hands with palms turned up in exasperation—these could become part of an emotional repertoire similar to the supply of epithets used by the bards. Alternately, interactors could be given a limited vocabulary of separate words or perhaps preformed phrases, like the bardic epithets, that could be assembled in many different ways. Such an artificial interface would work best if it is motivated by the situation, as, for example, in a visit to another country or in a conversation in restricted circumstances where things have to be said according to a strict protocol or secret code. But whether a writer creates the set of primitives out of gestures or phrases or a combination of the two, the challenge will be to make them as transparently expressive of emotion and intention as joystick fighting and link-navigation are in the current environment.

The next level of patterning after the stock phrase in the bardic storytelling method is what Lord refers to as the theme, that is, a generic narrative unit that can be fit into multiple narratives, a unit such as the departure of a hero, the catalog of ships, the dressing of a hero for battle, the boast of a hero before battle, and the death of a hero. The theme functions like a scene in a play or a chapter in a book. This is the key unit of segmentation that the poet focused on when memorizing a new story from the recital of another bard. Like the folksinger who concentrates on the chord sequence and the general rhyme scheme of the verses rather than on the exact words of a newly acquired song, the bardic performer did not think about line accuracy but focused instead on reproducing the order of the component themes. With no system of writing or recording with which to compare two renditions, bardic performers repeated the poem with perhaps only 70 percent accuracy of words but with exactly the same order of themes.

The plot events in electronic games and MUDs closely resemble these epic themes, because they draw their material from genres like fantasy, science fiction, and comic book heroics that are very close to the folktale tradition. The more filmic CD-ROMs rely on later formulaic genres such as the murder mystery or the horror film. Genre fiction is appropriate for electronic narrative because it scripts the interactor. When I begin a CD-ROM murder mystery, I know I am supposed to question all the characters I meet about what they were doing at the time of the murder and keep track of all the suspects' alibis. I will use whatever primitives I am given (navigation through the space, conducting an interview, picking up pieces of evidence and looking at them under a microscope, etc.) for enacting these prescripted scenes. In a Western adventure I can be counted on to try to shoot at the bad guys, and in a horror story I will always enter the haunted house. I perform these actions not because I have read a rule book but because I have been prepared to do so by exposure to thousands of stories that follow these patterns.

A mature narrative tradition will take advantage of this common base of formulas to refine the scripts, to offer the interactor a richer range of behaviors. For instance, since within the conventions of a mystery story it is already customary to send the detective to a mystery bar, CD-ROMs already include bars and the interrogation of bartenders. A refinement of this convention might mean offering well-lit booths and dark corners for the detective to choose from. A well-lit table might be safer in a confrontation, whereas a dark corner might invite more revealing disclosures. Or we might discover in a more domestic story that bringing breakfast to a lover or a box of crayons to a child will deepen a relationship and move the plot forward. Patterned activities like these could grow into new thematic units (like making friends or winning trust or showing loyalty) in new genres of electronic stories that focus on textured relationships rather than on puzzle solving and gunfights.

At the highest level of organization, Lord's bardic singers assembled their thematic units into plots. One very common pattern for the Yugoslavian poets was the story of the hero's return, which exhibited both constant and variable elements. The most constant part of the story was the return itself, which always included thematic units detailing disguise, deception, and recognition. Usually the return was preceded by an account of the hero's release from prison and was followed by an account of his return to prison to rescue someone else; often it included a recollection of a much earlier event in which the hero was summoned to war on his wedding day. The wedding story might be told as the beginning of the whole tale or just as a flashback from later events. Sometimes the prison part of the story did not end in the hero's release but in a refusal of release, followed by rescue of the hero by someone else. Sometimes the rescue tale elements were rearranged into a story of a bride being rescued from the enemy.

Lord realized that all the stories that center on rescue and release intertwined with marriage and battle were "basically one song" with many different plot possibilities. Both listeners and performers were constantly aware of other narrative possibilities growing out of the same thematic elements. Therefore, whenever the singer reached a thematic event that belonged to multiple story patterns, he would be "drawn in one direction or another" by "similarities with related groups" of songs:

> The intensity of that pull may differ from performance to performance, but it is always there and the singer always relives that tense moment. Even though the pattern of the song he intends to sing is set early in the performance, forces moving in other directions will still be

felt at critical junctures, simply because the theme involved can lead in more than one path.[5]

The singer's "tense moment" is a consciousness of a branching point in the formulaic composition process. Lord points out that the *Odyssey*, which is a similar "return" story, includes significant vestigial elements of another potential plot—in which Telemachus starts to set off to rescue his father—showing that Homer also felt this pull. Although the written tradition is based on a fixed set of events, the oral tradition is much more "fluid" and is based on what Lord calls a multiform story.

> Unlike the oral poet, we are not accustomed to thinking in terms of fluidity [of text]. We find it difficult to grasp something that is multiform. It seems to us necessary to construct an ideal text or to seek an original, and we remain dissatisfied with an ever-changing phenomenon. I believe that once we know the facts of oral composition we must cease trying to find an original of any traditional song. From one point of view each performance is an original. From another point of view it is impossible to retrace the work of generations of singers to that moment when some singer first sang a particular song.[6]

The bardic system is fundamentally conservative; it serves to transmit a fixed story from teller to teller and from generation to generation. But what it conserves is not a single particular performance but the underlying patterns from which the bards can create multiple varied performances. Their success in combining the satisfactions of a coherent plot with the pleasures of endless variation is therefore a provocative model of what we might hope to achieve in cyberspace. To do so we must reconceptualize authorship, in the same way Lord did, and think of it not as the inscribing of a fixed written text but as the invention and arrangement of the expressive patterns that constitute a multiform story.

Vladimir Propp and the Bardic Algorithm

Oral composition can even provide us with a specific algorithm for producing multiform stories. Around the same time that Milman Parry first began to notice the oral character of Greek epic, the Russian formalist Vladimir Propp set out to analyze a body of Russian oral narrative in order to arrive at a "morphology of the folktale." He was quite successful in his efforts at reducing a seeming "labyrinth of the tale's multiformity" to an "amazing uniformity." The set of 450 fairy tales Propp studied, though very different from one another on the surface, resolved themselves into variants of a single core tale composed of twenty-five basic "functions," or plot events, which we can think of as Propp's essential morphemes.[7] (Propp gave each of these morphemes its own symbol, which appears in parentheses to the left of each example below.)

After a brief introductory section (with its own distinctive morphemes, including the following: a family member absents himself from home, an interdiction is violated, a villain attempts reconnaissance, the hero is deceived, and so on), the first element of the story is one of these:

(A)   The villain causes harm or injury to a member of the family (nineteen variants ranging from threats to abduction and murder).

(a)   One member of the family either lacks something or desires something.

These elements are familiar to us from other fairy tales, even if we have never heard the Russian ones Propp studied. Here are some of the key story elements and their symbols:

(↑)   The hero leaves home.

(D)   The hero is tested (ten variants ranging from requests for help to challenges to fight).

(F)   The hero acquires a magical agent (a magic hen, a magic horse, magic foods, etc.).

(H)   The hero and the villain join in direct combat.

(I)   The hero defeats the villain.

(J)   The hero is branded.

(K)   The initial misfortune or lack is "liquidated," or resolved (eleven variants).

(L)   A false hero presents unfounded claims.

(M)   The hero is given a difficult task.

(N)   The hero successfully performs the task.

($\downarrow$)   The hero returns.

(Ex)   The false hero is exposed.

(U)   The villain is punished.

(W)   The hero is married and ascends the throne.

In addition to identifying the elements, Propp tried to establish the rules by which these morphemes are combined. He found that many morphemes came in related pairs, for example, the establishment of a misfortune/lack and its liquidation, the pursuit of the hero and his rescue, the introduction of a false hero and his exposure. Propp also noticed that the order of the elements in a story seems to be constant, even though any particular version of the story might lack some of the elements. For instance, the test of the hero always occurs after he leaves home and before he acquires the magical agent. In a particular version of the story, the test might be left out, but it would not be transposed. Like a cook who analyzes a dish to understand how it was prepared, Vladimir Propp has arrived at a recipe for the Russian folktale.

More than that, Propp's analysis makes clear that the formulaic underpinning makes folktales more intricate; it allows storytellers to weave together multiple different story sequences without becoming confused. For instance, a hero might receive more than one magical gift in the course of a story, a variation that would require a repetition of the elements that cluster around the gift pattern; or a second hero could be introduced to go on his own quest, forming a second story sequence that would be embedded within the first. Propp's abstract notation allowed him to chart such complex story structures. For instance, here is his representation of a story that grows out of two acts of villainy, each of which is resolved separately:

A14 I_____K9
A2   II_____K1

Here the villain commits abduction (A2) and murder (A14). The first part of the plot resolves the murder by reviving the dead person (K9). The second part of the plot resolves the abduction when the hero completes a search by an act of cleverness (K1). Propp's notation also reveals some of the rules for creating more divergent variants while still keeping the same basic story line. For instance, he noticed that the "struggle/victor" pattern (elements H-I) could be substituted for a "difficult tasks" pattern (M-N). When he finished analyzing all the extant tales, Propp was able to summarize all the variants of the Russian folktale in one inclusive representation. His work suggests that satisfying stories can be generated by substituting and rearranging formulaic units according to rules as precise as a mathematical formula.[8]

The Computer as Storyteller

Propp's algorithm is much more complex than most electronic games currently on the market but considerably less complex than current attempts to model stories in computer science laboratories.

The story line in most gaming software can be described in terms of two or three morphemes (fight bad guy, solve puzzle, die). MUDs also rely on the repetition of a narrow set of plot actions, often limited to combat, negotiation and ceremonial events. Indeed, the lack of plot progression in MUDs is an advantage, since a limited repertoire of stereotyped activities makes for more easily sustained role-playing. Adventure and puzzle games usually provide only one route through various game levels, which results in a very linear story despite the high degree of participation activity. Games that do offer choice-points leading to variant plot events are usually constructed with only shallow detours off the main spine of the plot. This is because even a story of less than a dozen branch points, with only two choices at each branching, would require hundreds of endings. Any branching story interesting enough to sustain our attention would therefore be too dense and confusing to write, since writers would have to work their way down each branch separately.[9]

Games that do provide narrative variety often do so through a simple substitution system. Just as one "magic helper" can replace another in a Russian fairy tale, so too can one hero replace another in a fighting game, often changing the emotional tone along with the joystick moves. But games do not allow substitution of thematic plot elements (e.g., a heroic labor instead of a struggle with a villain) the way fairy tales do. Games are limited to very rigid plotlines because they do not have an abstract representation of the story structure that would allow them to distinguish between a particular instantiation and a generic morpheme. That is, "level two" of a fighting game always refers to the same configuration, not to a set of rules by which it can be constructed. A morphological approach (a generic "level two") would require more ambitious programming but it would offer much greater plot variation; it would give the writer the power to tell the system how to generate variants without having to make each possible version individually.

Several kinds of abstract schema have been proposed by computer scientists as ways of representing stories, many of them based on a model of story structure grounded in cognitive theory. Most of these systems, however, have an unnervingly reductive quality to the humanist. For instance, Patrick Winston's analogy-making program Macbeth summarizes the plot like this:

This is a story about Macbeth, Lady Macbeth, Duncan, and Macduff. Macbeth is an evil noble. Lady Macbeth is a greedy, ambitious woman. Duncan is a king. Macduff is a noble. Lady Macbeth persuades Macbeth to want to be king because she is greedy. She is able to influence him because he is married to her and because he is weak. Macbeth murders Duncan with a knife. Macbeth murders Duncan because Macbeth wants to be king and because Macbeth is evil. Lady Macbeth kills herself. Macduff is angry. He kills Macbeth because Macbeth murdered Duncan and because Macduff is loyal to Duncan.[10]

This summary certainly fits Forster's definition of a plot as "a narrative of events, the emphasis falling on causality," but it is a sledgehammer causality that few literary critics or psychologists would accept because of its disturbing flattening of motivation. Of course, Winston is not trying to understand the play as a literary critic or a psychologist or to be moved by it as a member of the audience. He wants to use it as a pattern from which to predict behavior in similar situations, as a way of mimicking human reasoning by analogy in complex situations. For instance, here is a story that Winston's system finds analogous to Macbeth:

This is a story about Linda and Dick. Linda is a woman and Dick is a man. Dick is married to Linda. Dick is weak because he is drained. He is drained because Linda is domineering.[11]

Stereotypical thinking is both useful and pernicious. It is useful because it is a form of abstraction that helps us to organize information. It is pernicious because it distorts the world and can make it hard to see things individually. I had always felt that Lord's Serbo-Croatian bards were a wonderful cultural treasure until I reread his account after the recent war in the region and was chilled to come upon his matter-of-fact mention of the "willing" abduction of women from the enemy as just another variant of the rescue theme. It is important to remember that any abstract story system ultimately refers to the sorrows and pleasures of human life and that the story of any event depends heavily on who is doing the telling. A storytelling system that further calcifies the distortions of stereotypical thinking would be as destructive as the most bigoted and bloodthirsty bard. We humans already do enough mechanical thinking without enlisting machines to help us.

Furthermore, stories told from an abstract representation of narrative patterns but without a writer's relish for specific material can be incoherent. Here, for instance, is a computer-generated fable based on a representation of a plot as the solving of a problem or the attainment of a goal:

> Joe Bear was hungry. He asked Irving Bird where some honey was. Irving refused to tell him, so Joe offered to bring him a worm if he'd tell him where some honey was. Irving agreed. But Joe didn't know where any worms were, so he asked Irving, who refused to say. So Joe offered to bring him a worm if he'd tell him where a worm was. Irving agreed. But Joe didn't know where any worms were, so he asked Irving, who refused to say. So Joe offered to bring him a worm if he'd tell him where a worm was. . . .[12]

The program goes into a loop because it does not know enough about the world to give Joe Bear any better alternatives. The plot structure is too abstract to limit Joe Bear's actions to sequences that make sense.

Several ambitious proposals have been made to ensure coherence in computer-generated narrative by creating plot controllers capable of making intelligent decisions about narrative syntax on the basis of aesthetic values. Brenda Laurel, who sees the computer as an inherently theatrical environment, has proposed an interactive fiction system presided over by a playwright who would shape the experience into the rising and falling arc of classical drama.[13] Marie-Laure Ryan has proposed a story-generation system, derived from narratology theories, that would shape satisfying tales exhibiting symmetry, suspense, and repetition.[14] The challenge of all such ambitious schemes is in giving the computer enough knowledge of the story elements to decide what constitutes an Aristotelian recognition scene or a suspense-generating event.

One way of avoiding the arduous task of teaching the computer to understand the world well enough to make such aesthetic judgments is to code very specific story elements in terms of their dramatic function. Michael Lebowitz has created a storytelling system along these lines with morphemic segments derived from the staples of daytime soap opera stories, namely, amnesia, murder threats, forced marriages, and adultery. In Lebowitz's Universe system, the automated author is assigned goals, and the system then looks for fragments that will achieve those goals.[15] For instance, one commonly invoked goal is "keeping lovers apart," which could be satisfied by "lover drafted by the army" and "get partner involved with someone else." Like a soap opera author, the program tries to maximize plot fragments so that they serve the purposes of multiple stories. Here is an example of a story sequence generated with this program:

> Liz was married to Tony. Neither loved the other, and, indeed, Liz was in love with Neil. However, unknown to either Tony or Neil, Stephano, Tony's father, who wanted Liz to produce a grandson for him, threatened Liz that if she left Tony he would kill Neil. Liz told Neil that she did not love him, that she was still in love with Tony, and that he should forget about her. Eventually, Neil was convinced and he married Marie. Later, when Liz was finally free from Tony (because Stephano had died), Neil was not free to marry her and their trouble went on.[16]

Why is the story of Liz and Tony so much more engaging and coherent than the story of Joe Bear? Both are formulaic and machine generated, but Universe starts off with story fragments that are much closer to a writerly source and with a much more particularized representation of a plot. But Universe is also a limited model, since it does not provide for a participatory narrative or for a plot that comes to an end.

In fact, very few efforts have yet been made to create a system that accommodates both interactivity and directed plot. Although Brenda Laurel has been urging the development of such systems since the 1980s, they have received very little attention so far, perhaps because they would require closer collaboration between writers and computer scientists than has been possible up to now. The most promising work in this area has been done by the Oz group, at Carnegie Mellon University, led by Joseph Bates. The Oz group is attempting to create a system that a writer could use to tell stories that would include an interactor, a story world with its own objects, computer-based characters who act autonomously, and a story controller that would shape the experience from the perspective of the interactor.

The Oz group has modeled this system as a live-action theater game in which an interactor is placed in a threatening situation in a bus station.[17] Acting students improvised the parts of the computer-based characters, receiving instructions through headsets from an off-stage director who watched the action closely and set off events at appropriate times. The climax of the story is the moment in which the interactor is offered a gun and must choose whether to use it to protect a blind man from a thug, or to escape onto a departing bus. To an observer of the experiment (which was captured on videotape), the action seems painfully slow and the climactic moment rather chaotic. But to the interactor the scene was quite gripping; the offer of the gun was a difficult moral choice, a self-defining moment, because it seemed to be happening in real time. The work of the Oz group suggests that plot satisfaction in an interactive environment is very different from plot satisfaction in an audience situation. In order to ensure the appropriate dramatic pacing, we may need to develop story controllers that monitor all the elements of the environment, adjusting the fictional world with the same precision and forethought as a chess master choosing among complex strategies.

The complexity of pattern manipulation made possible by the computer seems to be pushing stories into the realm of higher degrees of abstraction and variation. But in pursuing complexity and abstraction, we run the risk of incoherence. Since the success of any abstract representation of plot will depend on how much control remains in the hands of the human author, we may find that less computational abstraction will produce more satisfying stories. Or we may discover new abstraction models that are closer to the way writers like to make up stories than the models that have arisen so far from the collaboration between cognitive theorists and computer scientists.

The Shaping Role of the Human Storyteller

Of course, the pleasure of storytelling lies not in the raw formulas but in the particularizing details. No one would curl up at night with a tale made out of a recitation of Propp's abstract morphemes. But, as my Russian-born students regularly remind me, the fairy tales Propp is talking about are beloved childhood memories, part of a rich culture of stories that are savored and delighted in to this day. We can glimpse what is so appealing about them in Propp's lists of variants and subvariants. For instance, here are some of the ways in which the morpheme of "the rescue of the hero" appears in the tales themselves:[18]

1. The hero is carried away through the air (e.g., fleeing at the speed of lightning, riding away on a flying horse, flying away on the back of a goose).
2. The hero flees, placing obstacles in the path of his pursuer (e.g., a human hero throws a brush, a comb, a towel, which turn into mountains, forests, lakes, or a superhuman hero tears up mountains and oak trees, placing them in the path of a she-dragon).
3. The hero, while in flight, changes into objects that make him unrecognizable (e.g., a princess turns herself and the prince into a well and dipper, a church and priest).

It is only when we read these enticing specifics that we feel the enchantment of the stories. We do not want to read about a generic Joe Bear even in a fable; we want to read about characters who have assimilated details that are vividly imagined and specific to their stories. For this we need the shaping presence of an author.

A story is an act of interpretation of the world, rooted in the particular perceptions and feelings of the writer. There is no mechanical way to substitute for this and no reason to want to do so. Our question instead should be, How can we make this powerful new medium for multiform narrative as expressive of the writer's voice as is the printed page? The answer is in coming up with strategies for giving the author direct control over all the many levels of artistic choice. The author must be able to specify all the elements of the abstract structure: the primitives of participation (how an interactor moves, acts, converses); the segmentation of the story into themes or morphemes (the kinds of encounters, challenges, etc., that make up the building blocks of the story); and the rules for assembling the plot (when events happen and to whom). The author must also be able to control the particulars of the story: all the substitution elements (instances of character types, dangers, rewards, places, travel experience, etc.) and all the ways in which each instance will vary. (Can we have a violent confrontation in a scene that includes the hero's mother? What kind of house will each of the possible villains live in?) We have only begun to think about how a writer would go about creating a story world out of such elaborate patterned elements.

Suppose we were to make a "cyberdrama" set in 1940s Casablanca, not an interactive version of the movie but a new narrative experience that uses some of the familiar adventure genre motifs that the movie employs so powerfully. The object would be to offer the interactor an opportunity to have many different adventures, by assuming the role of several distinct characters, all of whom are pursuing their own destinies in the French-controlled colonial city during World War II. We could begin by building a representation of the city itself as an immersive environment, including several nightclubs, with illegal gambling in a back room and a private office for the manager; some seedy hotels with dingy guest rooms; and an outdoor market with vendors' stalls. We might also create taxicabs, a bank, some private homes, and other sites not seen in the movie but necessary to sustain a sense of a real place. In our ideal system we would invent abstract representations of all of these spatial story elements, but we would only create specific ones each time we ran the story. We would also establish some unchanging venues, such as a police station and an airport and perhaps a particular café called Rick's. For the purposes of our new story, however, Rick might be out of town. Or perhaps we would show the events of the movie happening in the background. The familiar events could serve as a kind of timekeeper for the unfolding of each new story and would also make the fictional world more externally defined and therefore more likely to evoke our immersive trance.

We would then have to establish clear primitives of participation. For example, the interactor would be able to buy food and drink, walk and take taxis, touch things and people, and engage in dialogue. We would have some particularly important decisions to make in determining how to use language. There would be a trade-off between the variability of the plot and the extensiveness of the dialogue, since we will want to specify the dialogue separately for every possible interaction. If some of the dialogue is in text (words on the screen rather than prerecorded audio), we would have more freedom to vary it for each separate story, since we could assemble combinations on the fly and without having to record each possible combination in advance with the right dramatic inflection. But even if we could find a way to extend the computer-based character's dialogue, we would still have to limit the interactor's input so we could maintain control of the story. Perhaps we would make the interactor's character a non-English speaker and supply the character with a fixed vocabulary (perhaps in the form of a phrase book), which would require us to anticipate all possible utterances. Since the film has established the improbable premise that everyone is speaking English and since it contains characters who are using such a phrase book, this might feel dramatically appropriate.

We would then decide on the thematic units or morphemes of the story. Among the most likely would be: arrival in Casablanca (or establishment of story role if the character is a long-term resident), offer of letters of transit, offer of sexual encounter, meeting with the SS, offer of resistance activities, discovery of letters of transit, bargaining for letters of transit, decision between two lovers, confrontation with French police, evasion of the SS, and act of political resistance. These events would be built up from the basic formulaic elements of touching, taking, moving, speaking, and so on, and they would fit together in a prespecified, multiform way, like events in a Russian fairy tale, to form a number of coherent plots. We might want to include multiple confrontations with the French police and the SS for every character and to make sure these happen very early in the narrative experience to motivate the protagonist's need to leave and to resolve the situation. We might also want to introduce the former lover very early in the story, but we would not want every story to take the same shape. Instead, we would always include some summons to a secret meeting. Sometimes this summons would come in a note and sometimes in a phone call or a whispered word on the street. The meeting would always involve someone with whom the protagonist has an ambivalent relationship, not just a former lover but also, perhaps, an old friend who once acted in a crisis, a brother or sister who once betrayed a confidence, or a former enemy who is now fighting for a cause the protagonist believes in. Providing such characters to interact with the protagonist in the secret meeting would reflect our interpretation of the heart of the Casablanca story as the tension between ambivalent intimate relationships and larger moral impertives.

In setting up the plot we would have to decide on what kind of ending stories would have in this world. Since the danger of Casablanca is not merely death but corruption, the story choices should lead to consequences that are serious on many levels. Some

choices might lead to death, but the death might be noble and satisfying if it entails opposition to the Nazis. Other choices might lead to
escape at the cost of moral degradation. For instance, if you, as interactor, were to betray a character like Victor, it might not lead to suicide (because suicide would violate the interactor's sense of agency)
but it could lead to the Ilse character finding out and leaving you. Or,
in a less gamelike ending you could find yourself sitting at a table with
a newspaper report of the death lying in front of you next to a bottle
and glass. You would be able to pour the liquor and raise the glass but
not get up from the table. This enforced immobility would suggest the
despair of a person about to drink himself to death.

The more freedom the interactor feels, the more powerful the
sense of plot. Since plot is a function of causality, it is crucial to reinforce the sense that the interactor's choices have led to the events of
the story. It is common to talk about the physics of a simulated world,
that is, the way visual objects move, whether there are two or three
dimensions in the underlying representation, whether there is gravity,
friction, and so on. Stories have to have an equivalent "moral
physics," which indicates what consequences attach to actions, who
is rewarded, who is punished, how fair the world is. By moral physics
I mean not only right and wrong but also what kinds of stories make
sense in this world, how bad a loss characters are allowed to suffer,
and what weight is attached to those losses. In the game world the
moral physics is very slight; there are no real choices between good
and evil and no consequences to the horrific violence displayed. In a
story the moral physics must be more substantial and lifelike. We may
be satisfied with stories in which luck is against us if they make sense
in the imaginative world. For example, if we are being chased down
by Nazis in Casablanca, we may not necessarily expect to survive.
The stories in which characters are unsuccessful will add resonance
to the ones in which they are.

By generating multiple stories that look very different on the
surface but that derive from the same underlying moral physics, an
author-directed cyberdrama could offer an encyclopedic fictional

world whose possibilities would only be exhausted at the point of the interactor's saturation with the core conflict.[19] The plots would have coherence not from the artificial intelligence of the machine but from the conscious selection, juxtaposition, and arrangement of elements by the author for whom the procedural power of the computer makes it merely a new kind of performance instrument.

The Coming Cyberbard

The narrative invention and story coordination necessary for creating a cyberdrama would be considerable but it might be no more daunting than writing a serial Victorian novel or serving for several seasons as head writer of a multithreaded television drama—provided that computer scientists come up with the appropriate authoring tools. Since the writer's task is analogous to composing a multi-instrument musical performance, what is needed is a system for specifying story motifs that is as precise as musical notation and that works something like the packages now available for arranging music, that is, by letting the author specify one part at a time and then try out the combinations and make appropriate adjustments. We are a long way from having such a tool right now, but we are beginning to see some glimmers of how it might be constructed.

One way of giving the author control over both the abstract and the particular elements is to describe all the story elements as a system of interconnected "frames." The frame is a powerful conceptual format of the digital representation of qualitative information. It was first proposed by Marvin Minsky, the influential and controversial pioneer of artificial intelligence whose career has been devoted to representing human consciousness in terms of computational structures.[20] Minsky envisions human memory as a set of frames, each of which has "slots" or "terminals" in it. We can think of these frames as multisided blocks with Lego-like connectors of various shapes and sizes. Some of the information stored in the terminals would be specific instances of the item for which the frame is a pure abstraction, a kind of platonic ideal. For instance, "Grauman's Chinese Theater" would be stored as the specific instance of a movie palace of the 1930s, and "Roy Rogers" would be stored as an instance of "singing cowboy." Other slots would accommodate attributes of the item or procedures for using it, any of which might be a frame in itself. So one slot of the "movie palace" frame might be "glamorous features," which in the case of Grauman's Chinese Theater would include "handprints of the stars in concrete outside." The "singing cowboy" frame might include a slot for a theme song like "Happy Trails" and one for Roy's horse Trigger.

A frame is a good way for specifying formulaic structures, like a hero and his attributes, or elements of a murder mystery. It is also a good way of transcending rote formulas, since it allows multiple particularized representations of the same object or event. For instance, we might have a "toaster frame" in our heads with entries for different toasters we are accustomed to using and for their way of operating and peculiarities. Some of the things we do with toasters (like putting slices of bread into them) might derive from our general "toaster frame" information, and some, like not using bagels with the narrow-opening toaster at our aunt's house, might be specific to individual toasters, part of the quirky specifics of life that give stories authenticity. An advantage of frames is that we can always create a new instance when we meet a new item in the original category (another movie palace or another toaster). We can also subsume frames under other categories (e.g., movie theaters under entertainment venues and toasters under appliances), and we can share information in different frames by sharing terminals (e.g., we can understand that, despite their differences, a nineteenth-century opera house has cultural uses in common with Loew's Paradise movie house in the Bronx and that a toaster oven has things in common with a pop-up toaster).

We can give an item in a frame a state (the theater is open or closed, the toaster is working or broken) and make decisions based on that state without having to respecify everything we know about the general category and the particular object in order to make that decision. We can also store with our knowledge of a particular item the instructions for how to interact with it (e.g., a script for entering a movie theater or using a toaster). An item described in frame representation could also have different attributes depending on the other frames it belonged to; for example, all instances of toasters in appliance stores would have price tags no matter what their other toaster-like qualities, since they also have the attributes related to the frame "store merchandise."

The utility of frames as a memory structure is that a relevant frame is called up whenever some of its terminals are filled, allowing us to avoid having to explicitly observe everything about our world every time we perceive it. For instance, since we already have lots of information about our own kitchen and about kitchens in general filed in the appropriate frames in our mind, we do not have to reconstruct the concept from scratch when we enter someone else's kitchen for the first time. We know there will be a sink somewhere and how a refrigerator door opens and what sorts of things are likely to be on the counter or in the drawers or cupboards. This is, of course, exactly the sort of information that human beings bring to bear on their experience and that computers do not.

It would not be realistic to expect to be able to provide a computer with all the relevant frames about all the objects in a fictional world. We could spend a lifetime just describing toasters and teacups, since human common sense contains so much experiential data about even the most prosaic objects and events in the actual world. And even if we could fill the encyclopedic capacity of the computer with all this information, it would not make for more expressive digital narrative because stories do not draw on generic information about the world but from inflected interpretations of a purposely limited slice of human experience. Therefore, the computer scientist's use of frames as tools for representing the objective world and for reproducing the redundant connections of human reasoning is probably a dead end for creating satisfying narrative.

But suppose we attempted to use the powerful abstraction tool of the frame to represent not the infinitely describable real world but the very limited domains of genre fiction. For instance, we could develop a frame-based representation of the fictional world of the American Western. It might have a generic frame for a saloon specifying all the objects that are found in a movie saloon, the kinds of things that can happen there, and how those things are transacted, but it would not have to have any knowledge of the dimensions of the space in the saloon or of the kinds of liquor served unless such specifics were important to the narrative. It might, however, include very particular details about the kind of mirror over the bar and the circumstances under which it could reflect action elsewhere and the way people around it would react if it were smashed. A saloon might also have a bartender (who would have a frame of his own), dance hall girls (ranging on a scale from scantily dressed waitresses to prostitutes), gamblers, chairs (with instructions for breaking, as in movie sets), gunfighters, swinging doors, and so on. Among the events that could happen in a bar would be card games, cheating, accusations of cheating, and gunfights. A writer would specify all of these events generically in terms of their dramatically significant elements. For instance, there would be no need to teach the computer to play poker but only to know when to let an extra ace drop out of a sleeve.

The writer would create not only frames to represent all the possible thematic morphemes of the genre but also plot frames to specify all the ways they could be arranged for a single interactor. These frames might include a "mode" terminal containing substitution rules that would allow the same generic elements to be assembled in very different styles. A TV Western system, for example, might be able to run in *Gunsmoke* mode, *Maverick* mode, or *Annie Oakley* mode by varying the degree of violence allowed and the range of supporting characters and subplots. The same abstraction of a Western town might appear as three different fictional universes to different interactors, whose story preferences (as indicated by their first few choices, perhaps) would dictate the appropriate mode of operation. For instance, a player who stepped up to the bar and ordered

hard liquor while twirling his six-shooter might become involved in a
Gunsmoke-style fight that would end in a citizen's death and the escape of the killer, whom the marshal would then be asked to track
down. A player who dressed up fancy and sat down to gamble might
experience a comic saloon fight and many opportunities to outsmart
the aggressor and avoid a fight. A player who started with target practice and stopped outside the saloon to help an old man cross the
street might be offered a saloon fight in which no one would get hurt
but the villains could be subdued by trick shooting. All of these
scenes would have common elements—bad guy, citizen, insult, challenge—from the generic Western fight frame.

A frame-based authoring system would allow a writer to enter
each element in its generic and particular forms and would keep
track of the connections between them, for example, which kinds of
characters fit into which kinds of events. It would allow the writer to
cycle through the possible plot possibilities, eliminating many of them
and specifying appropriate choices or priorities for situations where
the story pulls in multiple directions. The off-the-shelf elements
would provide most of the story for a purely formulaic writer, but they
would provide only the palette for a more inventive one. As in any
genre, the more original the writer the more she would have to invent her own elements and the more actively she would inflect the
conventional formulas. What the computer would provide would be
a means for using formulaic patterning, in much the same way the
oral bards did, as a system for assembling multiform plots. The electronic system might be able to generate more variants than the author could ever read in a lifetime (let alone write individually), but
since she would have specified all the important details and all the
rules of variation, the computer would be merely the instrument of
the author, an extension of her memory and narrating voice.

Digital plot making, like other aspects of the medium, is still in an
incunabular stage. The technological resources of the gamemakers
are directed toward rapidly transforming visuals rather than expressive storytelling. The self-conscious webs of the postmodernists and

the link-happy exhibitionism of the Web soaps send us hopping from screen to screen in search of a coherent story. The more filmic CD-ROMs offer more extended story segments but embed them in a shallow branching structure that frustrates our desire for participation and agency. The MUDs offer extensive opportunities for participation in formulaic narrative environments, but the collectively generated stories are diffuse and repetitive. None of these formats puts the processing power of the computer directly into the hands of the writer. The experiments of the computer lab point to the possibility of much more powerful narrative tools, but they are still very remote from the storyteller's desire to simply enchant us or to grab us by the collar and tell us something more real than reality. Only when these disparate efforts begin to converge will the medium come into its own as an expressive art form. It seems to me quite possible that a future digital Homer will arise who combines literary ambition, a connection with a wide audience, and computational expertise. But for now we have to listen very, very carefully to hear, amid the cacophony of cyberspace, the first fumbling chords of the awakening bard.

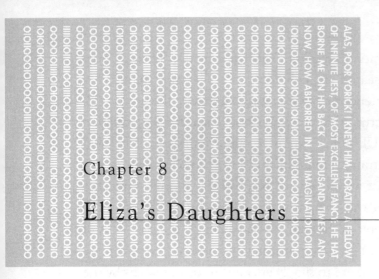

Chapter 8

Eliza's Daughters

The novelist . . . makes up a number of word-masses roughly describing himself . . . gives them names and sex, assigns them plausible gestures, and causes them to speak by the use of inverted commas, and perhaps to behave consistently. These word-masses are his characters.

—E. M. Forster, *Aspects of the Novel*

In any literary medium characters are illusions. Emma Bovary, David Copperfield, and Huckleberry Finn are "word masses," as Forster reminds us, and when they are translated to the movie screen, they only exist as montages of camera shots, pasted snippets of light and sound. What difference will it make, then, to create characters from bits, from digitized words, images, sounds, and—most significantly—from instructions for behavior? When Joseph Weizenbaum invented the computer-based character Eliza in 1966, he gave us a hint of the answer. Since then Eliza has had many direct and indirect descendants, enough so that we can identify some distinct technical and artistic strategies for making computer-based characters.

Chatterbots

Probably the most famous of Eliza's daughters is the virtuoso character known as Julia, developed by Michael Mauldin of Carnegie Mellon University. Julia is a "chatterbot," a text-based character like Eliza
who carries on conversations with the people around her.[1] Julia was
built to live on MUDs, and she has many agreeable social behaviors:
she plays the card game hearts, keeps track of other inhabitants, re-
lays messages, remembers things, and gossips. In short, Julia is good
company. Moreover, her physical presence is as real as anyone else's;
that is, she can hold things, perform actions like other MUDders, and
move from place to place. Like any other being, she appears in only
one room at a time. MIT researcher Leonard Foner, who has studied
her behavior appreciatively, records her dramatic appearance in answer to the typed-in command "page julia":

> You sent your summons to Julia.
>   Julia pages from Stevi's Kitchen: "I'm on my way to you, Lenny."
>   Julia is briefly visible through the mist.

Julia is clearly a chatterbot who knows how to make an entrance.
Foner also offers this example of her entertaining conversational
skills:

> You say, "Hi, Julia. What do you do?"
>   Julia says, "I gossip, Lenny."
>   You whisper, "Julia, tell me some gossip," to Julia.
>   Julia whispers, "I once heard baba say, 'Drop kick me, Jesus,
> through the goal posts of life.' "

Julia's responsiveness and appropriateness to her surroundings induces people to collaborate with her in little dramatic scenes. For instance, Foner describes his own delight in seeing her after she had
been off-line for a while (he immediately burst into the first line of
her theme song):

Lenny: Julia, Julia, she's our guide!

Julia: She directs us far and wide!

Lenny: Are you lost, can't find your way! Julia's here to save the day!

Julia bows.

Lenny laughs.

As a character with a female name who lives on MUDs, Julia usually engages in that familiar form of collaborative comedy known as pickup banter. MUDs only accentuate this already highly patterned and predictable behavior, making them the perfect venue for Eliza-like creatures, who excel at parodic interactions:

Space-Ace: So why won't you kiss me, Julia?

Julia: Sorry, I'm not into geeks, Space-Ace.

Space-Ace: Julia, I'm not a geek.

Julia: Then lose the pocket protector, Space-Ace.

Space-Ace: Julia, I don't have a pocket protector.

Julia: Congratulations.

Space-Ace: So kiss me, Julia.

Julia: I'd rather be overhauled by a Yugo mechanic, Space-Ace.

Space-Ace: Julia, kiss me.

Julia: Why, did your sheep leave you?

Space-Ace: Julia, kiss me.

Julia: Take a long walk through an unlinked exit, Space-Ace.

Julia's impersonation of an actual female MUDder was so successful that Mauldin was distressed to discover that one poor soul had spent thirteen days trying to seduce her into going with him to a private room for virtual sex. Among their more memorable interactions are these exchanges:

Barry: Are you a virgin?

Julia: As far as you're concerned, Barry, I always will be.

Barry: I'm inviting you to enjoy my private room.
Julia: So?
Barry: Well, I thought you liked me and wanted to be together.
Julia: I live for your approval, Barry.[2]

In some ways Julia is a female impersonator like the other female personae operated by male MUDders, and her character shows some of the exaggeration of a drag character. For instance, one of Mauldin's playful details was to program Julia to complain at appropriate intervals about her period and to make jokes about PMS. Here is one such exchange, elicited by one of Foner's colleagues:

Robyn: Julia, are you human?
Julia: Is the Pope Polish?
Robyn: Yes, Julia, but are you human?
Julia winks at Robyn.
Robyn: Julia, are you a computer?
Julia is offended at Robyn's suggestion.
Robyn: Julia, are you a woman?
Julia: I am female.
Robyn: And, Julia, are you human?
Julia: Not when I'm having my period, Robyn.
Robyn laughs.

Julia clearly gets most of her vitality from her comic shtick, but she also functions as an expert conversational partner on a limited number of topics. In this regard she is a kind of "query system," which is a form of artificial intelligence program that takes in users' questions ("When is the next bus to Cleveland?" or "What is the best form of birth control?" or "Is Lenny on the MUD today?") and consults a database to generate the appropriate answer. In order to make query systems succeed, one must limit their domain of expertise and then anticipate the many ways in which questions might be asked. Julia functions as a very simple query system in her ability to answer questions and sustain conversations about hockey or pets. She was given fluency in these topics in order to compete in a Turing test contest. The Turing test was proposed in 1950 by cryptographer Alan Turing as a way to judge if machines have matched human intelligence.[3] In its usual form, judges converse with both machine-based characters and humans through a keyboard and then decide which is which. When Eliza fooled people into thinking she was an actual person, many believed it would not be long before machines could pass sustained tests of this kind. In 1991, Hugh Loebner offered a $100,000 prize for the first program to pass the test. The prize has yet to be awarded, but the annual Loebner competitions are an entertaining reminder of the goal. Here is an excerpt from one of Julia's entries:

> *Judge:* Do you know anything about miniature schnauzers?
>
> *Julia:* I don't know anything about miniature schnauzers, Judge.
>
> *Judge:* Well, I have one. We call him Pucci. He is a crazy cat—have you ever had a cat?
>
> *Julia:* What a coincidence! I have two cats.
>
> *Judge:* I prefer dogs myself. Cats are too independent and they always jump up on things. By the way, are you a hep cat?
>
> *Julia:* People don't own cats . . . it's the other way around.
>
> *Judge:* How do you feel about declawing?
>
> *Julia:* What kinds of pets did you have as a child?
>
> *Judge:* Won't you answer my question? How do you feel about declawing?
>
> *Julia:* Guinea pigs are about as smart as cucumbers.

Julia is not a very believable character in the Loebner arena, even though she is able to fool people like Barry in the MUD. It is easier to create a fictional personality than it is to model knowledge of the world. In fact, Julia is more imaginatively present when she is less intelligent, because her sense of presence does not come from giving factually correct information but from demonstrating dramatically appropriate behavior. The MUD provides a social framework in which her formulaic responses make sense. Indeed, her exaggerated gender behavior is itself a good distracter: it makes people focus on a possible male-as-female impersonation rather than on the machine-as-human impersonation.[4] But what most brings Julia to life on the MUD is the fact that her interlocutors are also in character. They come to the conversation predisposed toward dramatic collaboration. A conversation with a chatterbot is a kind of improvised skit between human and computer-controlled actors. A successful chatterbot author must therefore script the interactor as well as the program, must establish a dramatic framework in which the human interactor knows what kinds of things to say and is genuinely curious about how the chatterbot might respond.

Authorship of Chatterbots

For a nonprogrammer the challenge of writing a character like Julia is to distract people away from the machinery, to make the illusion more interesting than the technical wizardry. One development that would help make such characters more compelling would be to make it easier for nonprogrammers to write them. To that end I have tried to give students in my interactive-fiction-writing course the opportunity to create their own chatterbots with a very simple authoring system. Although their results are only ambitious class projects and by no means complete characters, the evolution of their writing techniques offers a good case study of literary technique rising to meet a newly available technology.

One of the first things students learned in creating their own chatterbots was how to select a good character. The most successful characters have been those who are self-absorbed, evasive, or obsessive in familiar ways. For instance, the first set of characters the students created included several politicians in the situation of a press conference (it was an election year). Other good scenarios included a "babysitting job from hell," a folksy grandpa who loves eating gruel and going fishing, and a smug, pretentious artist. There were many comical

romantic partners, including a particularly revolting Continental se-
ducer and a "psychotic girlfriend," who was one of the most successful characters at sustaining a coherent conversation.[5]

Like all of the characters, Girlfriend begins with an opening line
that immediately establishes the situation and scripts the interactor:

> *Girlfriend:* So, why do you want to break up with me?
> *Interactor:* I need space.
> *Girlfriend:* Let's talk about the other woman.
> *Interactor:* What other woman? I've been totally faithful.
> *Girlfriend:* Where did you meet her?
> *Interactor:* Meet who? I told you I'm the faithful one in this rela-
> tionship.
> *Girlfriend:* How could you do this to me?
> *Interactor:* You did this to yourself, babe!
> *Girlfriend:* Do you love me?
> *Interactor:* Well, I used to. I know better now.
> *Girlfriend:* I can never feel for another man as much as I feel for
> you.
> *Interactor:* Well, maybe I'm being too hard on you.
> *Girlfriend:* How could you break up with me when you know
> how I feel about you?
> *Interactor:* You're smothering me. I need space.
> *Girlfriend:* You've hurt me deeply.
> *Interactor:* You hurt me too.

The rigidity of the computer is well used in the tunnel vision of the
character. The interactor enters into the collaboration immediately
and moves from a simple antagonistic attitude ("You did this to yourself, babe!") to opening up to her pain ("Well, maybe I'm being too
hard on you"), which gives the scene a hint of dramatic development.

For any chatterbot, the test of coherency is how it deals with the
inevitable problem that arises when the interactor's utterance contains no key word. One way to handle this is to do what Julia does in

the "declawing" example: start up a new topic (like guinea pigs). Another way is to sustain the previous topic, as Girlfriend does when she follows up "Let's talk about the other woman" with "Where did you meet her?" The most common strategy is to provide a set of "default utterances," comments that can fit in anywhere ("How could you do this to me?"). After this default strategy had been established practice for a while, one student used it to construct a salesman character who returns to his sales patter if confused. Salesman was the first character to use a default file as a means of providing a plot for the scene. Salesman begins with hard-sell optimism like this:

- My name, by the way, is Brian. Have you heard of the WidgetMaster product line?
- The WidgetMaster is currently in use in more than 10,000 homes across the nation!
- Let me know at any time if you would like to see a demo of the WidgetMaster in action.
- Might you be interested in purchasing a WidgetMaster today?

He then progresses through increasingly anxious self-disclosure:

- Please buy one! It really will change your life.
- I really need to hit my quota this month, too. Please?
- I don't want to pull a sob story on you here, but please, buy one?
- Look, ever since Mr. Widget's son took over, I've been under the gun.

Salesman ends in outright despair:

- Sigh.
- I give up. I'm going to miss my quota and lose my job.
- I can't believe my life has turned out like this.
- Sigh. Sorry for taking your time.
- I really don't know what to do.[6]

The final reply is coded as repeatable, so that the conversation from then on is peppered by the refrain.

In creating Salesman, the author invented a method of making the rules for a conversation serve as a dramatic structure for an interactive scene. This modest exercise is an example of how new narrative techniques can develop when a tradition of composition is supported by an authoring environment that does not require programming. If such software environments, created and refined by programmers working in collaboration with writers, were more widespread, chatterbots could move beyond their current status as engaging novelties. Computer characters who can carry on persuasive conversations could be an expressive narrative genre in themselves, as well as compelling elements in a larger fictional world.

Modeling the Inner Life

Although Joseph Weizenbaum chose the psychiatric interview only because it allowed him to bypass the problems of knowledge representation that haunt the Turing test encounters, other researchers of the 1960s and 1970s were drawn to the idea of the psychiatric interview in the sincere hope that the computer might serve as a therapist or as an aid to therapy through the modeling of the human psyche. Psychoanalyst Kenneth M. Colby, one of the most ambitious of these researchers, improved upon Weizenbaum's ELIZA by creating PARRY, a model of a paranoid person that was convincing enough to pass a kind of Turing test. Colby's description of his "hypothetical individual" reads like a doctor's presentation of a case:

> He is a 28-year-old single Protestant male who works as a stock clerk at Sears, a large department store. He has no siblings and lives alone, seldom seeing his parents. He is sensitive about his parents, his religion, and about sex. His hobby is gambling on horseracing, both at tracks and through bookies. A few months ago he became involved in a severe quarrel with a bookie, claiming the bookie did

not pay off a bet. After the quarrel, it occurred to him that bookies pay protection to the underworld and that this bookie might gain revenge by having him injured or killed by the Mafia. He is eager to tell his story and to get help in protecting him from the underworld. He is willing to answer questions about nonsensitive areas of his life and offers hints about his delusional system in an attempt to feel out the interviewer's attitude toward him.[7]

In fact, this person is as fully imagined as many fictional characters. Parry conversed via teletype with eight psychoanalysts, who also conversed in the same way with actual patients, some of whom were diagnosed as paranoid and some of whom were controls. Although the doctors could easily tell the paranoid patients from the nonparanoid people, they accepted Parry as another paranoid rather than as a mechanical impersonation. (This may say more about their self-fulfilling models of paranoia than about PARRY's computational subtlety, of course.) Here is a sample interview excerpt, as Colby reported it:

Doctor: How did you come to be in the hospital?
Patient: I am upset.
Doctor: How long have you been in the hospital?
Patient: About a week.
Doctor: In what way are you upset?
Patient: People make me nervous.
Doctor: In what way do they make you nervous?
Patient: People give me a funny look sometimes.
Doctor: Is there anything about people in particular?
Patient: Certain persons.
Doctor: Who are the certain persons and what is it that they do?
Patient: Do you know how the underworld operates?
Doctor: How do you mean?
Patient: Shouldn't I be?
Doctor: Why did you ask about the underworld?
Patient: What information do you have on the underworld?

Doctor: I don't know about the underworld but when you mentioned the underworld I thought you must have some contact with it and that's what I would like to know more about.

Patient: I try to avoid the underworld.

Doctor: Have you been actively trying to avoid the underworld?

Patient: No one has any power over gangsters.

At this point the psychiatrist remarks, "This response is a non sequitur. It's possible he's become so defensive about the underworld he's decided not to answer directly."[8] His explanation is a classic example of what computer scientists call the "Eliza effect," that is, attributing greater intelligence or intentionality to a machine than it possesses. It is also an example of the much older effect we earlier identified as the true source of Eliza's power—the human propensity to suspend disbelief in the presence of a persuasive dramatic presence. The psychiatrists were taken in by the persuasiveness of the animated presence; its real-time responses to their questions; and its very literally constructed, detailed obsessions with bookies and gangsters. Just as Julia's exaggerated female persona distracts from the man-or-machine question, Parry's paranoia focuses the observer on the quality of his mental life, not on whether it is authentic or not.[9] Colby sent transcripts of the interviews of Parry and of the paranoid patients to a group of randomly chosen psychiatrists and another group of randomly chosen computer scientists and asked them which was which, thus conducting a classic Turing test. The result was no better than chance in either group.

Parry could be taken as another example of a shtick comedy figure in the Eliza genre, but he is significantly different from Eliza in that Colby provided him not with a mere set of phrases but with a rudimentary model of an inner life. Colby gave his creature a "belief system" in which fairly innocuous beliefs about bookies lead to a core of emotionally charged delusions about gangsters. He gave Parry a model of a state of mind, including monitors for anger, fear, and mistrust. He instructed Parry to construct a model of his interviewer on the basis of each question and to decide if the interviewer's intent is malevolent, benevolent, or neutral. If Parry thinks a question is malevolent and his fear level is high enough, he clams up and stops talking about his delusions. He therefore has an inner state and a repertoire of behavior with which he can reflect it. Interactors can tell an "upset" Parry from a calm Parry, and the things they say can make a difference in the mood of this "patient."

Colby's most ambitious creation was an automated patient, Neurotic Woman, who hates her father because she feels he deserted her as a child but who censors this thought through standard Freudian processes. When she runs in "free-association mode," she is impelled to say "I hate my father," but she is also impelled to repress the thought, leading her to offer distorted expressions like "I hate Mrs. Smith (displacement) and "Father hates me" (projection). The Neurotic Woman program also runs as a psychoanalytic conversation in which the interactor plays the part of the analyst and can break into the free association now and then with carefully chosen suggestions. Here is an excerpt from this simulated analysis:[10]

> *Patient:* Father is indifferent to me.
> Mother says father abandoned us.
> Father preferred sister.
> I avoid father.
> Sister admired father.
> I feel tense now.
> *Therapist:* You admire father.
> *Patient:* No.
> I do not admire father.
> I admire Al.
> Mother ought not to admire Al.
> Mother detests father.

Colby was influenced by theories of cybernetics as well as by Freud in designing his model, and his program traces the patient's moment-to-moment state of mind (as expressed in numbers representing degrees of anxiety, excitation, pleasure, self-esteem, and well-being) through intricate feedback mechanisms that regulate the degree of distortion applied to any potential statement. The aim of the psycho-analyst is to come up with ideas that will move the patient toward confronting her contradictory and painful thoughts while raising her self-esteem and keeping her anxiety in check. Since the goal of the interaction is to enable the simulated patient to express her hatred of her father directly and thereby abandon her neurotic belief system, Colby can be credited as the first person to conceive of an automated fictional character with an inner self that is capable of change and growth.[11]

Although Colby's work was meant as science, it is more persuasive as literature. Both Parry and Neurotic Woman stand as tantalizing ex-amples of a computer-based character who is modeled not as a collec-tion of surface behaviors but as a layered personality.

Goal-Based Critters

As intriguing as automated patients are as literary models, they were abandoned by computer scientists in the 1970s and 1980s in favor of a behavioral and cognitive model of personality and social interaction that could be programmed in terms of "scripts, plans, and goals."[12] Experiments like Tailspin, which created the dithering Joe Bear we met in chapter 7, reflect a preference for describing characters not in terms of their psyches but their goals. In the 1980s and early 1990s, as computing power and available memory increased exponentially, computer scientists began exploring software and hardware strategies for "parallel processing," for creating systems that could do multiple things at the same time. Leading engineers turned from building all-encompassing centrally controlled systems to designing worlds made from a collection of "intelligent agents," each of whom was pursuing its own goals. This change in computer architecture has an equiva-lent effect for the creation of digital narrative. It is as if computer scientists stopped trying to build a world by coming up with an omniscient storyteller and decided instead to create it out of a collection of autonomous characters. Moreover, the characters they are currently building are not the single-minded, top-down planners of Joe Bear's generation. They are improvisers, aware of multiple goals at once and able to change their priorities and behaviors in response to changes in their environment. These characters are often called "intelligent agents."

Computer science research on "intelligent agents" often focuses on utility programs that act like servants. The classic example of a desirable software agent is one who goes out over the Internet and books airline flights and hotel reservations, consulting the client's preferences on budgeting and on arrival and departure times and taking advantage of opportunistic events like "frequent flyer" bonuses. Since the desire for such agents is often couched in terms very like those used by the early Disney cartoonists to create an "illusion of life,"[13] it is not surprising that interest in them has been closely intertwined with a desire to create fictional characters. For computer scientist Joseph Bates, the key goal for such research is "believability," which he has tried to achieve in several experimental character-based fictional worlds.

One of the first characters created by Bates' Oz group was Lyotard, the housecat. Lyotard was named for the postmodern theorist and is perhaps a fitting emblem of the times; he is the result of world-class expertise brought to bear on the problem of making a machine made of plastic and silicon impersonate a small, furry animal. It is a largely successful effort, one that could easily be translated to a virtual reality or graphics-based program, since Lyotard lives within a virtual house. For now, Lyotard's actions are expressed in text, as in the following representative scenario:

Interactor> Go to the dining room.
You are in the dining room.
A jar, a bread basket, a black cup and a large plate are on the end table.

Three bran muffins are in the bread basket.

Nine black sardines are in the jar.

Interactor> Take the jar.

You take the jar.

Interactor> Go south.

You are in the sunroom.

Lyotard goes to the sunroom.

Interactor> Give a sardine to Lyotard.

You offer the black sardine to Lyotard.

Lyotard runs to the dining room.

Interactor> Follow Lyotard.

You run to the dining room.

Lyotard looks around nervously.

Interactor> Pet Lyotard.

You pet Lyotard.

Lyotard bites you.

Interactor> Hit Lyotard.

You hit Lyotard.

Lyotard is now shaking.

You hear a meow.

Lyotard runs to the sunroom.

Interactor> Look around.

You look around yourself.

The end table and the small chair are in the dining room.

The bread basket, the black cup and the large plate are on the end table.

The three bran muffins are in the bread basket.

Lyotard looks around nervously.

Interactor> Eat a muffin.

You eat the bran muffin.

Lyotard goes to the dining room.

Interactor> Wait.

You wait.

Lyotard goes to the kitchen.

Interactor> Follow Lyotard.

       You go to the kitchen.

       Lyotard is in the kitchen.

       You hear a meow.

Interactor> Look around.

       You look around yourself.

       You are in the kitchen.

       A long counter, a cabinet, a beautifully crafted shelf, a plastic trash can, a yellow bowl, a red superball, a white toy mouse and Lyotard are in the kitchen.

       A small plate is on the beautifully crafted shelf.

       The red catfood is in the cabinet.

       The blue catfood is on the long counter.

       You hear a meow.

Interactor> Pour the jar in the bowl.

       You pour the jar in the yellow bowl.

       Lyotard eats a black sardine.

Interactor> Wait.

       You wait.

       Lyotard eats a black sardine.

Interactor> Wait.

       You wait.

       Lyotard eats a black sardine.

Interactor> Pet Lyotard.

       You pet Lyotard.

       Lyotard eats a black sardine.

Interactor> Again.

       You pet Lyotard.

       Lyotard looks lazily at you.[14]

Although this scene may seem far from great drama, it is a very promising achievement. With Lyotard, Bates and the Oz group have succeeded in creating a cat with "emotions, social behavior, and personality." They have done this with an architecture that gives him an inner life that is centered on goals (such as satisfying hunger and re-laxing) but that also includes a representation of his emotions. For in-stance, Lyotard begins by being fearful of the interactor and not very hungry, which is why he is not interested in the sardine and runs away from the interactor. When the interactor follows him from the sun-room to the dining room, Lyotard's goal of escaping fails, which makes him more aggressive. Therefore, when the interactor, whom the cat recognizes as an object of mild hate, goes to pet him, Lyotard forms the goal of "respond negatively to contact," which results in his choosing to bite the interactor. Why does Lyotard not hold a grudge? Because the interactor stays away from him, allowing his angry feel-ings to decay. Meanwhile, the cat's hunger is building. Therefore, by waiting for the cat to get hungry and by then offering him the special treat of sardines in a nonthreatening way, the interactor is able to make friends with him. This demo script is a satisfying dramatic scene, a collaborative improvisation much like a conversation with Eliza but based on gestures rather than words. But to what degree is Lyotard a characterization rather than a mere mechanized model of a cat?

Lyotard is certainly a complex machine. His psyche is built on a cognitive science schema that is widely used in the computer model-ing of personality.[15] His inner life is built on an intricate but precise calculus in which events are compared against goals, actions are com-pared against standards, and objects are compared against attitudes; Lyotard's psyche is a giant emotional algebra equation in which all the values are changing all the time. Success in a goal yields the emo-tion of joy and failure yields the emotion of sadness, but this schema can also represent ambivalence. Since even a simple housecat is a complex, dynamic system, many goals will be active at any given time, and some actions will satisfy one goal but frustrate another. As a behavioral and motivational model, Lyotard and his kind take us far beyond the "kill/don't kill" characters of action games. Lyotard can capture complex emotions like reproach, which is defined as the reaction to another person acting in violation of one's own inner standards; thus, Lyotard displays reproach when an interactor sits down in the cat's favorite chair. This framework is both elegant and absurd. It is elegant in that one can account for a wide range of emotions (including composite emotions like anger, which is represented as a combination of reproach and sadness) using a limited set of building blocks and for a range of emotional intensity that is expressed quantitatively (e.g., dislike of interactor = 1; dislike of dogs =10). But the cognitive model of emotions quickly becomes absurd when we try to apply it to the emotional states of actual human beings (dislike of Barney = 1; dislike of Hitler = 10), and it seems the very antithesis of what we value in literature, which is the careful examination of ambiguous situations open to multiple interpretations. A Tolstoy of the next century could hardly model Anna Karenina's conflict between her love for the passionate Vronsky and her love of her son by setting a panel of affect sliders and filling in a template with her goals and standards.

Moreover, even for the more modest task of describing a housecat with the same level of believability as a cartoon cat, the abstract science can only take us so far, for Lyotard's behavior in this scenario could not have been produced by merely mapping some common cat behaviors (eat when hungry, bite when angry) onto a standard emotion machine. Since their goal was believability, the authors were forced, almost against their scientific instincts, to give Lyotard some charm, to make him a particular character. For this goal they discovered that the cognitive science formulas did not take them far enough and that there was no scientific taxonomy to take them the next step of the way. Instead, they had to give Lyotard a set of features in addition to the canonical emotions, features that they present almost apologetically as invented without a known "structured set" and "very ad hoc." The authors' list of proposed features for Lyotard include many of the traits humans find most entertaining in housecats, namely, their capacity to be curious, content, aggressive, ignoring,

friendly, proud, and energetic. Without the aggressive feature, for instance, Lyotard would not have been disposed to bite the interactor in
the above scenario. In other words, Lyotard's most dramatically interesting behavior arises from a specific personality structure the authors
improvised, on top of the more generic model, just for him.[16]

The emotional schema is therefore more like a palette than a portrait. It provides a way of specifying elements of the personality and of
linking behavior to an interpretive model of the character's inner life.
The need for an ad hoc, unscientific characterization strategy is even
clearer in the more ambitious Oz group project called Edge of Intention, a multicharacter world in which oblong animated figures called
Woggles jump and slide around a two-dimensional landscape displayed on the computer screen. Here the researchers explicitly differ-
entiated the characters by mapping each of their emotions to a
personality-specific feature. For example, the same emotion of "fear"
is mapped to the "alarm" feature in the vulnerable Shrimp but to the
"aggression" feature in his nemesis, the bullying Wolf. But Oz designers still found it very hard to communicate these emotional states to
interactors, since the Woggles' world does not offer the interactor a
familiar script, as the housecat scenario does, to provide the dramatic
context for the characters' actions. In fact, the most expressive and
lifelike character trait arose from a programming glitch that caused
Shrimp to bang his head on the ground now and then in what Bates
considered a sort of nervous tic. At exhibits of the world of Woggles
around the country, this head banging stole the scene. The researchers who knew this behavior to be incoherent in Shrimp's true
emotional life were struck by how it immediately attracted people's
attention. "To our surprise," wrote Bates, "they build psychological
theories, always incorrect, about Shrimp's mental state and seem to
find him much more interesting and alive for having this behavior."
Having created their most expressive gesture by accident, the computer scientists were left to debate the extent to which believability
rests on clean abstraction and how much it rests on quirks.[17]

Lyotard and Shrimp are a promising beginning for computer-based characters with both coherent behavior and believable personalities. The emotional abstraction system, like any other formalization of human personality, can be reductive and clumsy, and many would dismiss it out of hand as a persuasive representation of our inner lives. However, it can provide us with useful building blocks for creating patterns of intelligible behavior or stencils for drawing in the general outlines of character. Nevertheless, for the work of characterization, for making creatures that engage us, we need the free-form drawing, not the stencil; we need the ad hoc element, the features that turn a block drawing into a face. We need to place the formulaic elements, the stenciled images, within an idiosyncratic arrangement based not on science but on an interpretation of the world. We have to build the cat to reflect how humans feel about their housecats rather than what they might learn about them in the dissecting room. If we are to reach the point where we can create complex computer-based characters with the same expressiveness that is true of the characters we now create with word masses and shots of film, then we must start by building authoring systems that will put the abstraction expertise of the precise computer scientist into the interpretive hands of the quirky artist.

Multicharacter Environments

We are just beginning to have experience with fictional environments in which automated characters interact with one another. For instance, at MIT and elsewhere Eliza runs within an environment that also includes a repository of the sayings of the syndicated cartoon figure Zippy, the pointy-headed clown whose surreal adventures make fun of the extremes of the popular culture.[18] This led to the inevitable meeting of the two. If one types in the command "psychoanalyze-pinhead," Zippy's sayings are fed to Eliza, and though he cannot listen to her, she responds to him in the same way she would to any interlocutor. Most of their exchanges are nonsensical, but amid pages of uninteresting non sequiturs come coherent exchanges like this:

> *Eliza:* I am the psychotherapist. Please, describe your problems.
>
> *Zippy:* We are now enjoying total mutual interaction in an imaginary hot tub.
>
> *Eliza:* Why do you say that?
>
> *Zippy:* I don't know WHY I said that. I think it came from the FILLINGS in my rear molars.
>
> *Eliza:* What makes you believe in your rear molars?
>
> *Zippy:* Give them RADAR-GUIDED SKEE-BALL LANES and VELVEETA BURRITOS!!
>
> *Eliza:* Maybe your plans have something to do with this.
>
> *Zippy:* Thank God!! It's HENNY YOUNGMAN.

When Eliza asks, "Why do you say that?" and Zippy replies, "I don't know WHY I said that," it is just a lucky accident. But the surreal unflappability of Eliza is a perfect foil for Zippy's outrageous non sequiturs.[19]

If we were to create a world for Zippy and Eliza to share we would want to maximize such exchanges. We would have to create them so that they each knew in advance the kinds of things the other was likely to say and had a repertoire of ready responses. Even putting aside the difficulties of getting a computer to understand language, this is a formidable task. Yet computer scientists are moving ever closer to multicharacter environments and are beginning to experiment with making coherent activity from the independent actions of multiple agents.

For instance, interactors with the Edge of Intention exhibit were allowed to operate one of the Woggles, like a videogame character, using a mouse to make it hop and slide through the simple landscape. The interactor's Woggle can greet and be greeted by the other creatures, influence the dynamic between the bullying Wolf and the timid Shrimp, and even lure other Woggles into a game of follow the leader. One of the problems of this arrangement is that it is "very difficult for people interacting with the creatures to stay aware of what is happening," because too many things are going on at once without the benefit of staging aimed at focusing attention on a central action. In other words, in addition to performing one's own character's repertoire of actions, deploying them appropriately and responding to the other characters in a multicharacter world, one must have a way of synchronizing these individual actions with the general action so that one is presented with a coherent picture. And if there is more than one interactor, the staging problem has to be solved either collectively or individually for all of them.

One solution to the staging problem of multicharacter improvisations is found in the practices of the commedia dell'arte, the popular Italian theater tradition that flourished throughout Europe from the Renaissance through the eighteenth century and that has influenced theater traditions from opera to *Saturday Night Live*. The commedia was based on a handful of stock characters, exaggerated types associated with particular actors in the troupe. The basic characters were the same whatever the plot, just as John Wayne or Groucho Marx or the battling lovers played by Spencer Tracy and Katharine Hepburn were the same from one movie to the next. A small number of actors (usually no more than seven) were able to perform a large repertoire of plays, ranging from farces to heavy melodramas, even though none of the plays had a written script. They did so by developing predictable formulas of interaction that gave shape to their improvisations.

Instead of a script, the actors relied on a scenario that offered clear entrances and exits and a paraphrase of each scene. For instance:

Pollicinella enters with a lantern and a sword. He is awaiting his master. He lies down, dowses his light and makes ready for sleep, whereupon

Don Giovanni leaps from the balcony. At the noise Pollicinella wakes up. They play a scene of combat in the darkness; then they recognize one another and both leave to make ready for their departure for Castile.[20]

Once the director of the company gave them the scenario, which told them when to go on and what their goal was for each scene, the actors relied on their stock characters to fill out the illusion. Whatever the setting or the story, there would always be two old men, two zany servants, a pair of lovers, a witty confidante for the prima donna to talk to.

How did the actors synchronize their utterances? In part they relied on memorized bits and pieces that were taken from the written tradition (and therefore quite polished) but appropriate for many occasions, such as elaborate insults, romantic verses, or a refused lover's complaint. In addition, they had rituals of interaction that they could adapt to varied content; for instance, the mistaken identity fight presented above could be reused in multiple plays. There were also formal patterns to the dialogue, as in this excerpt from a lovers' quarrel:

> *He:* Go! . . .
> *She:* Disappear! . . .
> *He:* . . . from my eyes.
> *She:* . . . from my sight.
> *He:* Fury with the face of Heaven.
> *She:* Demon with mask of love.
> *He:* I curse . . .
> *She:* I shudder . . .
> *He:* . . . the day that I set eyes on you.
> *She:* . . . at the thought that I ever adored you.
> *He:* How can you dare . . .
> *She:* Have you the insolence . . .
> *He:* . . . to look at me again?
> *She:* . . . to remain in my presence?

And so on through parallel accusations, denials, and renewed vows of devotion.[21] The actors also had stock comic stage business, called *lazzi,* such as the following:

Lazzo of the shoes:

When they are about to take Pulcinella to prison he says he must first tie his shoelaces. Then he bends down again, grabs the legs of his two guards, throws them down and runs away.

Lazzo of the fly:

Pulcinella, having been left by his master to guard the house, on being asked if there is anyone inside, replies that there isn't even a fly. The master discovers three men there and reproaches Pulcinella, who replies: "You didn't find any flies, you only found men."

Lazzo shut up:

While his master is talking Pulcinella is continually interrupting. Three times his master tells him to shut up. Then when he calls for Pulcinella, the latter pays him back in the same coin and says "Shut up!"[22]

The actors were able to improvise their parts because they worked within these formal patterns, and, indeed, current improvisational comedians are trained with very similar techniques.

With this degree of formulaic patterning in mind, it becomes possible to think of generating scenes between procedurally described characters. Certainly the battling lovers would be easy to do, as would the mischievous servant and impatient master. To make such scenes work, it would be less important to model the characters' emotional state than to give them the right patterns of interaction, namely, the *lazzi*-like formulas that would let them anticipate one another's remarks and respond appropriately.

Pulling the Strings of the Digital Puppet

The insides of a digital character should perhaps resemble the improvisational materials of an actor—including set speeches, stage business, and plot patterns—more than the insides of an ordinary person, with emotions, beliefs, and superego. Parry's specific paranoid thoughts bring him to life in a way that his having knowledge, for instance, of how to eat with a fork would not improve. Such an improvisational digital actor might be thought of as a kind of marionette whose dramatic presence depends partly upon things that are unchangeable (such as a painted-on face) and partly on behavior invented on the spot out of a repertoire of actions; just as the puppeteer improvises from the limited possible positions of the strings, so too might the procedural author program the actors to improvise by combining elements of their behavioral repertoire.

In fact, researchers at New York University have already created two charming animated actors, who, though not yet ready to perform in the commedia, can invent their own staging in a graceful and expressive manner. They have a Renaissance look to them and physiques reminiscent of Don Quixote and Sancho Panza. One is tall and thin, with graceful fingers and a mournful face, the other round with a heavier gait. Although they cannot speak, they gesture expressively and move about on their digital stage, improvising in pantomime as if waiting for a sluggard playwright to arise and script them.[23]

To what extent do we allow such actors to pull their own strings and to what extent do we put them in the hands of a plot controller? One way of thinking about how this might work is to look at another house pet, a distant cousin of Lyotard the cat. Silas T. Dog, created by Bruce Blumberg for the ALIVE Project at the MIT Media Lab, was designed as a potential actor in interactive narratives; he is largely autonomous, but he can also accept a director's commands at four levels of his complex inner life. Suppose Silas were playing with a little girl in the "magic mirror" VR environment (described in chapter 2) and the program decided to make the encounter more dramatic by getting Silas to do something mischievous, like snatching a virtual steak from a virtual kitchen table. The plot controller could prod him into action by sending a command to his motor system (*go forward*), to his behavior modules (*find a juicy steak*), or to his motivation system (*you are very hungry*). Moreover, the plot controller could change

Silas's perception of the environment in ways that are not visible to other interactors. For example, one way of getting Silas to pay attention to the little girl would be to attach a computer model of a dog biscuit to her hand that would be visible to Silas but not to her.[24]

Silas's multipart architecture, which is based on studies of animal behavior and is optimized so that he does not dither between conflicting goals, could just as well serve as an architecture for a character's personality. Instead of a motor system, we give the character a repertoire of actions appropriate to the story world (e.g., *send letter, change clothing, order murder of enemy, form posse*). On top of this layer of simple behaviors we could have motivational modules (*fall in love, swear revenge, seek allies*) and behavior modules drawn from Proppian morphemes (*defect to the enemy, woo the princess, search for the outlaws*). Instead of a virtual pet we would now have a virtual actor, ready to be sent onstage as the multiform plot calls on him. The device of the invisible dog biscuit would be risky for the creation of believable narrative, since it introduces the possibility of a fictional world that is cohabited by multiple beings who have no fixed common reality. However, giving each character its own private sentience of the world would be a virtue, since it would allow the plot controller to direct the discovery of important story elements (to make sure, for example, that the Giant does not see Jack hiding in his beanstalk home until just the right moment).

The strong point of Silas's architecture is that it allows him to respond to his environment spontaneously and to put together complicated strategies for doing things. In fact, characters like these are so complicated that they have the potential to walk away with the story altogether. They pose the question of just how autonomous we would want a fictional character to get.

Emergence as Animation

Autonomous agents like Silas and Lyotard offer the exciting possibility of what computer scientists call emergent behavior; they are able to act in ways that go beyond what they have been explicitly pro-grammed to do. For example, in the scenario described earlier, Ly-otard chose to bite the interactor from among a range of behaviors open to him at that moment. His creators had made him aggressive, and they had arranged his motivational priorities so that he would give a higher weight to responding to a nearby hand than to satisfying his general need to relax. But they did not specifically instruct him to bite that particular interactor at that particular moment. The action emerged from the intricate combination of sensations, emotions, and personality traits that shape Lyotard's simulated consciousness.

It is an important moment in human history to be able to make machines that exhibit emergence. It is a sign that we have reached a new threshold in our ability to represent complex systems—systems of any kind, whether thermodynamics, war strategies, or human be-havior. In the first cybernetic models, systems were thought of as being under a central command structure, like a thermostat, and computer programs were built in simple hierarchies with one master program that controlled other programs, or subroutines. Later sys-tems were often based on the notion of a "finite state automaton" that chugged from one complex state to another in sequences that could be charted in a neat map of circles connected by lines. But as our models of the world have become more complex, systems have become decentered: their processing operations are distributed among many entities, none of which is in central control, and the possible states of the system as a whole are no longer thought of as fi-nite. The new emergent systems have reached such a degree of intri-cacy that they are their own description; there is no other way to predict everything they are likely to do than to run them in every pos-sible configuration.

It is the focus of current work in computer science to try to control the unpredictable, to ensure that autonomous agents will always "do the right thing."[25] For instance, Bradley Rhodes in Pattie Maes's group at MIT, has created a Three Little Pigs cartoon world (dis-

played on a desktop computer) from creatures that are close cousins
to Silas T. Dog.[26] The protagonist of the story is Wolf, who is an autonomous agent with multiple ways of catching and eating a pig, such
as dynamiting the pig's house or using a pogo stick to hop to the roof
and down the chimney. He does not have any central program that
tells him how to do these things; his behavior is always improvised in
a moment-to-moment fashion and depends on which motivations are
strongest and on what his current situation and environment are like.
It is exhilarating to watch Wolf in action, because it is never clear
which strategy he will take as he responds to the scurryings of the pig,
the availability of his tools, and his adjustable set of goals. For instance when he feels both hungry and destructive, he walks to the
straw house, huffs and puffs till he blows it down, then walks over to
the pig, picks it up, and eats it—a very efficient Wolf. But when his
desire to be in high places is activated, instead of huffing and puffing
he goes over to the pogo stick, picks it up, pogo-bounces to the chimney of the house, climbs down the chimney, drops the pogo stick,
picks up the pig, eats it, picks up the pogo stick, and then bounces
back to the roof of the house—an exuberant and whimsical Wolf.
The character improvises each of these sequences out of his repertoire of abilities, just as we choose among various activities on the
basis of our mood. But it is also easy for Rhodes, as god of this environment, to render Wolf hopelessly confused by giving him conflicting goals of equal weight. If Wolf's desire to be in high places is set
equal to his hunger, Wolf may be too hungry to leave the pig and yet
too fixated on his cherished pogo stick to drop it long enough to eat
the pig—a comically neurotic Wolf. This is exactly the sort of situation that computer science aims at eliminating, yet it is exactly the
kind of situation literature relishes. From *Hamlet*, to *Anna Karenina*,
to yesterday's soap opera episode, it is dramatically depicted ambivalence rather than efficient goal selection that makes a sequence of
events into a story. Rather than eliminate such moments, a procedural author using Wolf might want to enhance them by giving Wolf

some expressive behavior, such as rushing back and forth between pig and pogo stick or banging his head like the Woggle Shrimp, to dramatize his absurd state of mind.

E. M. Forster thought there were two kinds of characters in fiction: "flat" ones, who perform their shtick in the same way throughout the narrative, and "round" ones, who can learn and grow. Eliza was a flat character, but Silas, Lyotard, and Wolf are—at least potentially—round ones. In digital narrative, the flatter the character the less risk it carries of breaking credibility. A character whose sole behavior is to shoot at the protagonist will always behave appropriately, but a character who can do more things—who can shoot, seek food, and elude police, for example—may dither or get lost in subgoals while ignoring events of larger importance. The more unpredictably emergent the behavior, the more it may need to be dramatically justified as a sign of a frantic or foolish person. Characters with emergent behavior may make much better keystone cops than romantic heroes.

Nevertheless, as Forster pointed out, we are more interested in characters who are capable of surprising us than in those who are flat and predictable. Forster praises Jane Austen, for instance, for creating characters who seem complex enough to live outside the limits of her plot:

> Suppose that Louisa Musgrove had broken her neck on the Cob. . . . The survivors would have reacted properly as soon as the corpse was carried away, they would have brought into view new sides of their character, and, though *Persuasion* would have been spoiled as a book, we should know more than we do about Captain Wentworth and Anne. All the Jane Austen characters are ready for an extended life, for a life which the scheme of her books seldom requires them to lead, and that is why they lead their actual lives so satisfactorily.[27]

Forster is making an argument for modeling characters not in terms of the plot demands but in terms of their whole inner life. But if we were to create such round characters on the computer, they might leave the scheme of the story altogether. How would we judge if an emergent behavior is satisfying? The same way Forster judges round-ness: "The test of a round character is whether it is capable of sur-prising in a convincing way. If it never surprises, it is flat. If it does not convince, it is flat pretending to be round. [A successful round char-acter] has the incalculability of life about it—life within the pages of a book."[28]

What we look for in a created character is not mere surprise but revelation. The unexpected behavior of a fictional character must be surprising in the way that human beings are surprising; it must tell us something we recognize as being true to life. Computer scientists often use the word *random* to describe the "incalculability of life," but characters who display randomly surprising behaviors are unconvinc-ing; they are merely flat characters pretending to be round. A truly round character would surprise the interactor by acting in a way that is consistent with its known behavior but that takes it to a new level. Its emergent behavior would have to come from a set of possibilities intentionally put there by the author, like a handful of seeds sown and then left to the vicissitudes of weather and environment to either flourish or die.

For instance, for a Casablanca simulation we might create a char-acter like the French woman at Rick's bar who is dating a German of-ficer but joins in the singing of "La Marseillaise" when Victor leads it.[29] Sabine (as we will call our character) would be given equal moti-vations of patriotism and expediency, as well as multiple, and possibly overlapping, states of mind (e.g., related to inebriation, infatuation, desire for economic security, and awareness of physical vulnerability) whose changing values would control how susceptible she might be to patriotic appeals at any given moment and also how likely she would be to display her patriotism if it were aroused. In a system in which the other characters and the major events are unpredictable, the procedural author would not know in advance whether Sabine would act as a collaborationist or a patriot in any particular scenario. Although her moral choices would be emergent behavior, the product of a particular moment and a particular history of experience within the story, the author would completely predetermine Sabine's innate capacity for such a moral conflict. To the interactor who meets her, Sabine would not appear to be acting in a random manner. Her behavior would be surprising but convincing, demonstrating that she had been created as a person with many possibilities, even with a kind of free will.

It remains to be seen whether we can capture the illusion of "the incalculability of life" with the emotional calculus of the computer. In the meantime, the most pleasurably surprising interactive characters may be those who are created with considerably less programming. For instance, the performance/installation artist Toni Dove has proposed an installation of "responsive cinema" in which two women characters, one from the past and the other from the future, are presented to a viewer whose movements are tracked with a motion sensor. As the viewer/interactor moves closer to each of the projected figures, she becomes more confiding. The process of navigating the installation is a social process, a growing intimacy or perhaps an enacted revulsion from one character or the other.[30] The content of what the characters say is conveyed in traditional dramatic materials—actors, a script, a filmed performance. No behavior emerges from such an installation that has not been carefully scripted, but the character will emerge to the interactor and the perceived sense of intimacy between them will grow as a result of their encounter.

To delineate characters with a capacity for revelation rather than computational emergence—the Anna Kareninas rather than the Silas T. Dogs—we will have to rely on moving images and written words rather than on programming code for the time being. But the next generation of graphics-based characters—animated figures capable of patterned conversations and a range of coherent gestures—should be much more fully alive than their predecessors. For instance, we can already purchase animated figures ("Dogz" and "Catz") to live on our computer screens that are similar to Silas and Lyotard in their spontaneity and computational design.[31] My own "computer pet," Buttons, is a frisky dog who romps happily when I open the program, races over to drink the water and gobble the food I put out for him, and leaps up to chase a virtual ball and bring it back to me. When I pet him, he bobs his head to direct my strokes, wags his tail, and growls with delight, and he blissfully goes off to sleep when I rub his belly. More than that, Buttons can learn tricks in somewhat the same way a real puppy can. If I hold up one of his digital treats, he will jump around and run through his repertoire of behaviors. If I give him the treat after he does the appropriate trick, he gradually learns to respond to the offer of bone-shaped digital dog biscuits by doing somersaults. Buttons, who has grown from a puppy to a larger dog since I installed him, has such a real presence for me that I sometimes feel guilty when I do not open the program and play with him. I find myself feeling proud of his affectionate personality, which is the result of the constant petting and good treatment I have given him. I know that the possibilities of his life are open and that if I had punished him with his spray bottle too often or in an arbitrary manner, Buttons's personality would be hostile and withdrawn. Playing with my computer pet is a pleasurable diversion, much more satisfying to me than stroking a stuffed animal or watching a cartoon. Characters like Buttons are a new kind of doubly animated figure, alive not only by the artfulness with which they have been made to look and move but also by the artfulness with which they have been programmed to be spontaneously responsive to the interactor's actions. The second product in the "computer pet" series, Catz, even incorporates this responsive behavior into a rudimentary story format by introducing another character—a mouse the cat can chase across the screen. The success of the Dogz and Catz programs may mark the beginning of a new narrative format, centered on appealing animated characters who might soon be as complex as Silas or Lyotard. Such characters could draw the interactor into more ambitious collaborative dramas.

Such modest incunabular creatures may seem hopelessly far from what we can achieve with Forster's "word masses," but they are nonetheless part of the same effort at understanding what it means to be human. Twentieth-century science has taught us that an important part of the answer to that question lies in understanding how complex systems like the ones the computer can embody for us resemble living things. In the centuries that have elapsed since the invention of the printing press, and as a result of the increased knowledge it has made possible, we have slipped further and further from our once cushy niche in the great chain of being—just above the conquered animals and below the encouraging angels. After Copernicus, Darwin, and Freud, we can no longer think of ourselves as somehow at the center of the universe, animated by the paternal finger of Michelangelo's God, or even (as we had hoped in the eighteenth century) as essentially innocent and rational creatures. Our solace until recently has been to celebrate our place in nature, our separateness from the increasingly mechanical world we are creating around us. Now, in the past few decades, that comforting thought is also being challenged. The brain scientists have speculated that consciousness itself may be understandable as an emergent phenomenon, the result of numerous unintelligent neurons all lighting up at just the right moment. As we slowly learn to model the processes of human thought and demystify them, the brain is left staring into a dizzying mirror. With oddly celebratory bravado, the computer scientist Marvin Minsky is fond of proclaiming that human brains, in fact, human beings altogether, are simply "meat machines." But if we are merely meat and merely machines, how are we to value ourselves and one another?

It has always been the job of the narrative imagination to answer such questions. In our time, part of the task of redefining what it means to be human lies in animating the machine, in using its system-modeling abilities to bring forth life—cuddly, affectionate, amusing, and recognizable—from empty matter. Digital dogs and cats invert the notion of a meat machine by turning an automaton into a pet. They make the idea of the mechanical less frightening by bringing it into our cultural space and domesticating it, just as our distant ancestors made the frightening world of the beasts less so by turning the wolf into a watchdog. We may not want to acknowledge a connection between ourselves and the mechanical world, but to be alive in our time is to be faced with this reflection, like it or not. We are compelled to search for the boundary, to find out what is left within us when we take away what we think of as "meat" or "machine." With the creation of these playful, quirky exploratory characters, then, the narrative imagination is beginning to awaken to this task.

PART IV

New Beauty, New Truth

Chapter 9

Digital TV and the Emerging Formats of Cyberdrama

Though the technology of the Star Trek holodeck remains improbably distant and the puzzle mazes, shooting games, and tangled Web sites of the mid-1990s have only begun to tap the expressive potential of the new medium, these first experiments in digital storytelling have aroused appetites, particularly among the young, for participatory stories that offer more complete immersion, more satisfying agency, and a more sustained involvement with a kaleidoscopic world. Although the tools of true procedural authorship are still in their early stages, it has become increasingly easy for interactors to construct their own worlds on the MUDs or to build custom game levels for open-architecture fighting games. For those who are not ready for procedural engagement, the preparation of digital text, audio, and video is increasingly accessible through off-the-shelf software. Web site design is fast becoming as easy as desktop publishing. Just as everyone who can cope with a keyboard and mouse can now make a greeting card, soon everyone who can master word processing will be able to design a simple Web page, complete with hyperlinks to other sites and color graphics.

As more and more people are growing as facile with digital environments as they are with pen and pencil, the World Wide Web is becoming a global autobiography project, a giant illustrated magazine of public opinion. Independent digital artists are using the Web as a global distribution system of underground art, including illustrated stories, animations, hypertext novels, and even short digital films. Although science fiction and fantasy narratives will always remain strong in cyberspace, the documentary elements of the Web—the family albums, travel diaries, and visual autobiographies of the current environment—are pushing digital narrative closer to the mainstream.

At the same time that legions of new Web surfers are busy debating politics or posting digitized pictures of the family schnauzer for the enjoyment of distant dog lovers, media conglomerates are trying to carve up cyberspace into revenue-producing fiefdoms. The entertainment industry has looked upon the world of bits as merely a new delivery channel, a simple wire for carrying their vast inventories of content to another market. They have been slow to understand what people look for in a digital environment, and they are likely to remain conservative in the creation of digital products, seeking only to modify the familiar formats of film and television so that they will somehow become interactive. The shape of narrative art and entertainment in the next few decades will be determined by the interplay of these two forces, that is, the more nimble, independent experimenters, who are comfortable with hypertext, procedural thinking, and virtual environments, and the giant conglomerates of the entertainment industry, who have vast resources and an established connection to mass audiences.

Looking ahead to the next forty years—the working life of the generation that has grown up with videogames and educational computing tools—we can expect a range of narrative formats to emerge as authors look for ways to preserve the customary pleasures of linear narrative while exploiting the essential properties of the digital medium with increasing sophistication. In this chapter we tear our eyes away from the distant horizon of the holodeck to focus on the entertainment products of the more immediate future. If the current multimedia CD-ROMs are the equivalent of the "photoplay," then what will be the next giant steps that will carry electronic narrative down the path from additive to expressive form?

The Hyperserial: TV Meets the Internet

One of the clearest trends determining the immediate future of digital narrative is the marriage between the television set and the computer. The technical merger is already under way. Personal computers marketed to college students allow them to switch off the central processing unit and tune in the latest episode of *Friends* on the same screen they use for word processing. The most computer-phobic couch potatoes can now buy a "Web TV" that will allow them to point and click their way across the Internet and even to send and receive e-mail, using an ordinary phone line. American television is rapidly moving toward a high-definition digital standard, which will turn the broadcast TV signal into just another form of computer data. Meanwhile, the Internet is beginning to function as an alternate broadcasting system; already it offers a wide assortment of live programming, including on-line typed interviews, digital radio programs, and even live video coverage of rock music concerts, club openings, and performance art. As television channels and the World Wide Web come closer together, the telephone, computer, and cable industries are racing to deliver the new digital content to the end user faster and in greater quantities. The merger that Nicholas Negroponte has long been predicting is upon us: the computer, television, and telephone are becoming a single home appliance.[1]

From the consumer's point of view, the activities of watching television and surfing the Internet are also merging, thus driving the marketplace to create new frameworks of participation. Television viewers populate hundreds of computer chat rooms and newsgroups, often logging on to these collective environments while watching the shows in order to share their responses with fellow audience members.

Broadcasters have experimented with displaying some of these comments in real time, as subtitles beneath the images of an entertainment program, as questions for interviewees, or as quotations at the beginning and end of news segments. The network formed by the cooperative venture between Microsoft and NBC exists as both a Web site and a cable television station; these two separate venues are so intertwined and mutually referencing that it would be hard to say which one is "the" MS/NBC. They are one entity, even though they now appear on two separate screens. Viewer digital participation is moving from sequential activities (watch, then interact), to simultaneous but separate activities (interact while watching), to a merged experience (watch and interact in the same environment). Although we cannot yet predict the economics of the television–Internet merger, these increasing levels of participatory viewership are preparing us for a near-future medium in which we will be able to point and click through different branches of a single TV program as easily as we now use the remote to surf from one channel to another.[2]

The more closely the new home digital medium is wedded to television, the more likely it will be that its major form of storytelling will be the serial drama. As we have already seen, the daytime soap opera has already been translated into the more participatory Web soap now popular on the Internet. Adding motion video to the format will increase demand for the dramatic immediacy and more tightly plotted action that we expect on TV. It will be hard for the chattily written Web soaps—which are based on a scrapbook metaphor—to compete in the same environment as television serials once the novelty of Web surfing has passed. At the same time, linear television will seem too passive once it is presented in a digital medium, where viewers expect to be able to move around at will.

Probably the first steps toward a new *hyperserial* format will be the close integration of a digital archive, such as a Web site, with a broadcast television program. Unlike the Web sites currently associated with conventional television programs, which are merely fancy publicity releases, an integrated digital archive would present virtual artifacts from the fictional world of the series, including not only diaries, photo albums, and telephone messages but also documents like birth certificates, legal briefs, or divorce papers. Such artifacts appear in the best of the current Web soaps but do not sustain our interest without the motivation of a central dramatic action.

The compelling spatial reality of the computer will also lead to virtual environments that are extensions of the fictional world. For instance, the admitting station seen in every episode of *ER* could be presented as a virtual space, allowing viewers to explore it and discover phone messages, patient files, and medical test results, all of which could be used to extend the current story line or provide hints of future developments. The doctors' lounge area could contain discarded newspapers with circled advertisements, indicating, for example, that Dr. Lewis is looking for an apartment in another state or that Dr. Benton is shopping for an engagement ring. An on-line, serially updated virtual environment would open up a broadcast story in the same way a film expands a story told in a stage play, by providing additional locations for dramatic action or wider coverage of the characters or events merely referred to in the broadcast series. We might see more of the home life of the *ER* doctors, perhaps noticing that Mark Green keeps a photo of the absent Susan Lewis next to a picture of his daughter or that Doug Ross has held on to the medical ID bracelet of a woman who died partly as a result of his out-of-control sexual life. Like the set design in a movie, a virtual set design would be an extension of the dialogue and dramatic action, deepening the immersive illusion of the story world.

All of these digital artifacts would be available on demand, in between episodes, so that viewers could experience a continuous sense of ongoing lives. A hyperserial might include daily postings of events in the major story line—another fight between feuding characters or a set of phone messages between separated lovers—that would be alluded to in the broadcast segments but detailed only in the on-line material. The Web-based material might also contain more substantial development of minor characters and story lines. Maybe Shep, whom Carol broke up with last year, is sending her letters telling her how he is dealing with the stresses of his job as an emergency medical worker, or perhaps the ex-prostitute with AIDS is in danger of losing her apartment. By filling out the holes in the dramatic narrative, holes that prevent viewers from fully believing in the characters, and by presenting situations that do not resolve themselves within the rhythms of series television, the hyperserial archive could extend the melodramatic broadcast drama into a more complex narrative world.

Putting broadcast television into digital form would also allow producers to make previously aired episodes available on demand. A hyperserial site would offer a complete digital library of the series, and these episodes, unlike the same content stored on a VCR tape, would be searchable by content. Viewers could call up individual segments of past episodes (the diner scene in which Mark finalizes his divorce agreement) or view a single continuous story thread (the deterioration of Mark's marriage) that was originally woven into several episodes. Such an encyclopedic representation of the complete series would offer television writers the larger, more novelistic canvas that serial drama has been moving toward for the last two decades. Writers could think of a hyperserial as a coherent, unfolding story whose viewers are able to keep track of longer plot arcs and a greater number of interconnected story threads. Compared to today's television writer, the cyberdramatist could explore the consequences of actions over longer periods of time and could create richer dramatic parallels, knowing that viewers would be likely to juxtapose events told months or even years apart.

There are several ways the complex organizational powers of the computer could be used to support a much denser and more demanding story world. William Faulkner was looking for a similar technological aid when he asked his publisher to use different colors for the print in *The Sound and the Fury* to guide the reader through Benjy's section of the story, a device that would make the time-jumping stream of consciousness of the mentally arrested boy comprehensible without the elaborate charts it otherwise requires from painstaking college professors. Faulkner also included a map of the town of Jefferson in the endpaper of *Absalom, Absalom!* that indicates where some of the events of his novels occurred, including not just the location of some of the more colorful murders but also of the pasture the Compson family sold so that Quentin could go to Harvard. The tongue-in-cheek map ("William Faulkner, Sole Owner and Proprietor") binds the multinovel, multifamily, multicentury saga together, giving us a taste of how Faulkner himself saw his mythical Yoknapatawpha County, not as a mere backdrop for his elaborately spoken stories but as a continuous geographical and historical realm that transcended all the stories told about it. The encyclopedic capacity of the computer allows for storytelling on the Faulknerian scale and invites writers to come up with similar contextualizing devices—color-coded paths, time lines, family trees, maps, clocks, calendars, and so on—to enable the viewer to grasp dense psychological and cultural spaces without becoming disoriented.

In the Victorian era, arguably the pinnacle of novel writing in English, fiction writers often published in weekly or monthly installments that would then be collected and rereleased in bound volumes. Cyberdramatists would be in a similar position and would have the same advantage of writing for two kinds of audiences—the actively engaged real-time viewers who must find suspense and satisfaction in each single episode and the more reflective long-term audience who look for coherent patterns in the story as a whole. But the digital storyteller would also be aware of a third audience: the navigational viewer who takes pleasure in following the connections between different parts of the story and in discovering multiple arrangements of the same material. For instance, a *Homicide* viewer might want to see more of how Pembleton's struggle to regain his mental functioning after a stroke has affected his relationship with his wife and infant. Or we might be offered a chance to get a fuller understanding of the nurse on *ER* whose ineptitude is a risk to patient safety but who is herself a victim of the hospital's policy of rotating senior nurses away from their areas of expertise. In a well-conceived hyperserial, all the minor characters would be potential protagonists of their own stories, thus providing alternate threads within the enlarged story web. The viewer would take pleasure in the ongoing juxtapositions, the inter- section of many different lives, and the presentation of the same event from multiple sensibilities and perspectives. The ending of a hyperserial would not be a single note, as in a standard adventure drama, but a resolving chord, the sensation of several overlapping viewpoints coming into focus.

Mobile Viewer Movies

The hyperserial model of cyberdrama described above is based on a transitional situation in which viewers are alternately watching tele- vision broadcasts and navigating a Web-like environment accessible from the same screen. But as digital television evolves as a delivery medium, viewers may find themselves unable to sit still for a conven- tionally told two-hour story. Just as the movie camera made the stage box seem too confining, so may the computer mouse make the direc- tor's camera seem too confining. Interactor/viewers may want to fol- low the actors out of frame, to look at things from multiple vantage points. We can already see evidence of such viewer restlessness in the hyperactive camera style of the most filmic television series (*Homi- cide, NYPD Blue*), in which the noncontinuous cuts and rapid circling movements of the often handheld cameras reflect the audience's own desire to roam around the space, to experience the action in three di- mensions, and to jump forward to the next interesting moment as quickly as possible. Although critics who are strongly attached to older forms of presentation might see such restlessness as evidence of a shortened attention span or an increased need for stimulation, it can also be seen as the expression of a more active curiosity or eager- ness to look around for oneself and make one's own discoveries. In some ways this desire to anticipate the next story move is similar to the impatience that comes just before individual literacy, when mem- bers of a culture (as in the early Renaissance) or a subculture (Victorian women) or an age group (grade-school children) can no longer stand to be read to but want to go through their own chosen reading at their own pace.[3]

To satisfy this desire, writers can create stories told in simultaneous actions, like the bedroom farce *The Norman Conquests* or the slice-of-life domestic drama "Evening," both discussed in chapter 6. Viewers would watch a "mobile viewer" cyberdrama, remote control device in hand, ready to click and branch through the story as it unfolds. The dramatic action would look like any ordinary television show, but whenever one character in a group of two or more exits to another room of a house or goes to another place in the fictional world, the viewer would have the option of choosing whom to follow. These choice-points need not look like computer program menus, nor should they stop the action in its tracks. Glorianna Davenport's Interactive Cinema Group at the MIT Media Lab has come up with several graceful alternate presentation styles in which a continuous movie plays before the viewer, offering automatic default sequences when no choice is made or responding to the suggestive positioning of a cursor by displaying an appropriate alternative selection in a non-interruptive, seamless manner.[4]

Moviegoers of the future may watch a single visual presentation but be offered multiple sound tracks. Everything that is said aloud in the scene might be on one track, available to everyone, but the private thoughts of various characters would be on their own tracks. A movie of a poker game or a sting operation might keep the motives of the protagonists secret from one another; since viewers could choose which character to align with, different members of the audience would watch the same scene with very different information. Moviegoers might be lured back to see such a movie again from a different point of view or to gain access to the thoughts of a character whose motivations were hidden from them the first time. Viewers in a 3-D theater watching a scene of an exotic café might hear all of the conversations spoken normally by people at their own table but would also be able to eavesdrop on whispered conversations or on people at adjoining tables by leaning their heads toward the speaker. This multidirectional audio, an enhancement of existing sound technology, would serve to make the perception of three-dimensional space much more concrete. Since these possibilities would lead to multiple viewings of the same film, and would therefore yield greater revenue for movie production companies without requiring the creation of additional footage, they seem likely to be attempted.

The mobile viewer approach could also be combined with the hyperserial. Perhaps a future *ER* will offer us a choice of trauma rooms in which to locate ourselves, or a future *Homicide* might offer us a choice of murder investigations to follow. Viewers who did not indicate explicit choices or who were watching with conventional television sets would see a continuous drama made up of default scenes, just as viewers with black-and-white sets were unable to take full advantage of the first color programs. But those with interactive access could choose to see more of some plotlines than others and could follow certain characters more closely. All of the dramas would end appropriately at the same time, and mobile viewers would also have a sense of having chosen from among several sequences to pursue the action or situation most dramatically intriguing to them.

This mobile viewer format would be very well suited to the current television genre of the problem drama, which addresses a socially charged issue, like racism or abortion, on which viewers hold very different views. A mobile viewer cyberdrama could be presented in such a way that viewers' choice of point of view would influence the kind of information they receive. Choosing to see the story in a particular way would therefore be a self-revealing act that might leave the viewer questioning his or her values.

The cyberdramatist would have the task of constantly arousing the mobile viewer's curiosity, fears, and sympathies, since every choice made by the mobile viewer should be expressive of a particular moment of imaginative engagement. Such choices, which would not correspond to a simple right–wrong dichotomy, should be interestingly different from one another and even more revealing when juxtaposed. One problem of delivering such a multibranching story over a television-like apparatus would be the conflicts it would engender over who controls the remote control device, who among the viewers in a household gets to choose the path of the story. It may become the custom for viewers to take turns in controlling the story narration, or perhaps broadcasters will offer mobile viewer stories several times within the same week so that viewers can experience multiple versions or so that the teenagers of the household can see the story from a different point of view than that of the adults or the women can see a story that differs from the one the men see. Any story that is presented in this way would have to offer the right balance of common and divergent experiences so that viewers are engaged by the same central situations but then see provocatively different versions of the events.

Mobile viewer audiences might then be offered the opportunity to converse with one another in chat rooms that are configured as sites within the universe of the program (i.e., sites such as cafés or squad rooms or school cafeterias). The treatment of controversial subjects in divergent narratives and the subsequent public on-line discussion would be a particularly appropriate format for television, which serves as a medium for what David Thorburn has called "consensus narrative," that is, for stories that define the concerns of the society and present the received wisdom about these concerns.[5] This format would provide a less voyeuristic way of engaging people in discussions about the kinds of disturbing behaviors upon which sensational talk programs focus. Issues of gender identity, sexual behavior, child rearing norms, or domestic violence could be framed by compelling stories that would then provoke discussion.

The creation of a commentary space within the fictional universe would put the viewer in the role of a member of a Greek chorus, a sounding board for the concerns of the protagonists.[6] The characters themselves (in avatars operated by the writers or by the improvising actors) might visit such an arena during a time of emotional crisis or dramatic choice and engage the viewers in on-line conversation.

Landview's Dorian (on *One Life to Live*), a particularly villainous character, could appear at an open house for fans in which they could be-
rate her for her evil scheming. Sympathetic viewers could visit a dying
character on *ER* to provide encouragement. A character on the verge
of an abortion or facing a child's disturbing sexual behavior or deciding whether to risk her job to reveal some corporate misdeed could
bring the dilemma into the shared fictional space for moderated discussion. An actress on *NYPD Blue* recently complained that people
on the street stop her to tell her that she—that is, her character—was
wrong to turn down a marriage proposal from the character played by
the handsome leading man. In an *NYPD Blue* hyperdrama, viewers
could come into a virtual squad room to gossip about the reluctant
bride's decision and to offer their opinions or to hear her venting her
ambivalent feelings. Such an engagement with a wider community
might make the writing on such shows more psychologically believable, particularly for the female characters, who are often not imagined with sufficient interiority to sustain viewer involvement.

The role of choral commentator would come easily to television
fans, but once they find themselves within the fictional world of the
series, they may not be content with just passing through and com-
menting on the story. They may want to move into the story world
and play a more active role in the plot.

Virtual Places and Fictional Neighborhoods

Currently the most inhabited fictional spaces—the MUDs—are
made of words alone, but as the Internet becomes faster and more capacious, as the conventions of three-dimensional environments become standardized, and as graphical authoring tools become more
functional and user-friendly, there will be an explosion of virtual architecture that will make the public digital environment look less like
a billboard-strewn highway and more like a populated landscape. In
the next decade, as the dungeons and forests of the MUDs are trans-

lated from words into three-dimensional images, more and more users may find themselves residing in such shared fantasy kingdoms.

Perhaps the first steps in this direction will be in the form of immersive visits to pleasurably explorable 3-D dreamscapes. The videogame manufacturers are already moving in this direction by offering worlds so well realized that the moat at the front portcullis is as appealing as the adventure within the castle. As joysticks and VR gear allow us greater mobility (not just up, down, left, and right but also in and out of a 3-D space) with more power of observation (i.e., with the ability to switch position as if we were operating a camera focused on the dramatic action) and with less physical encumbrance or need for manual dexterity, interactors will be lured into worlds where they float, tumble, and arc through thrillingly colored spaces, fly through imaginary clouds, and swim lazily across welcoming mountain ponds. The nightmare landscape of the fighting maze, in which we feel perpetually imperiled, may give way to enchanting worlds of increasingly refined visual delight that are populated by evocative fairy-tale creatures.

A visit to such a space will combine the rhythmic kinetic pleasures of dancing with the visual pleasures of sculpture and film; the space itself will be expressive, as our movement through it will be, and the landscape will be filled with objects of desire and enchantment. We will go out over digital networks to experience the thrill of entering previously inaccessible environments: an erupting volcano, a primeval rain forest, a distant planet; we will walk across the parted Red Sea with Moses or sit down to a performance in a virtual Elizabethan theater. These compelling immersive landscapes might constitute a new kind of pastoral art, an artificial re-creation of nostalgically fantasized natural or historical environments. Just as ancient Greek city dwellers enjoyed reciting verses about frolicking shepherds, so too will twenty-first-century citizens of the information age enjoy transmuting their data-laden screens into elfin groves, Victorian parks, or galactic fireworks displays.

We will go to these environments alone or with others, and perhaps in "smart costumes" like Laurel's crow body (see chapter 2). We will fly along with virtual geese and pet digital unicorns. The rooms will be richly textured and atmospheric. The creatures will be inherently charming in the way that children and small animals are charming to us. They will be easy to interact with because they will trigger our most basic interactive impulses—to offer them food, pet them, and clap with pleasure as we watch them cavort. Gradually these lushly realized places will turn from spectacle experiences to dramatic stages. We will move from the pleasures of immersion and navigational agency to increasingly active and transformational experiences.

Once we are accustomed to being on the virtual stage, there will be many routes into the world of cyberdrama. The exploratory spaces and chat rooms of television hyperserials will allow viewers to progress from playing the roles of explorer and choral commentator on the action to playing the familiar role of the television neighbor. From Ozzie and Harriet's home to Dan and Roseanne's, we have been virtual guests in scores of family living rooms, even as they have simultaneously been guests in ours. When the television set becomes digital, we will find it even easier to imagine ourselves within that virtual domain which lies somewhere between the television sound stage and our own living room. We will not, of course, move in with the characters on the screen, but we may very possibly move next door to them, occupying a contiguous virtual space and experiencing events, in persona, that are also happening to the characters in the series.

The Lucasfilm games are already moving in this direction. The first *Star Wars* arcade game allowed the player to repeat the actions of the movie hero, thus enhancing the player's enjoyment and excitement when events in the movie were duplicated in the game-play. To hear Alec Guinness's voice whisper "Use the Force!" was to become Luke Skywalker for a moment in a magical way. But the PC-based game

Rebel Assault is even more exciting because it allows players to have their own adventures, parallel to those in the movies and carefully woven into the same event sequence and time frame. In *Rebel Assault,* the player is not Luke himself but a rookie cadet who rises to squad leader and goes on all the key missions of the *Empire Strikes Back* but from a different part of the battlefield. For instance, at one point the player sees the movie heroes leave the ice planet Hath and is left behind to mop up and to escape as best he can while the Empire's forces close in. When my own son at age thirteen watched the movie again after mastering the complex videogame, he jumped up and down with excitement when he recognized the parallel sequence. "I was there!" he cried out. "I stayed on the planet after Han leaves. It was even more dangerous for me!" Like Don Quixote, he was able for that moment to act within a beloved narrative he had only witnessed before.

As 3-D environments become more detailed, children and adolescents will be increasingly drawn into virtual environments that function as satellites of the communities described in movies, comic books, and, most compellingly, broadcast television series. For instance, a program like *Dr. Quinn: Medicine Woman* might set up a series of virtual frontier towns some miles away from the series' Colorado Springs location, towns populated by interactors who could choose to be blacksmiths, barbers, general store owners, saloon keepers, scouts, and, of course, female doctors and who could be given their own homesteads or boardinghouse rooms in particular physical locations within the fictional world. The creators of the series could set some plots in motion within these towns and let other actions arise spontaneously from the role-playing of the participants. Area-wide events that happen on the series, such as an influenza epidemic or a confrontation with General Custer, would also happen in the neighboring towns. (Role-playing interactors could find themselves informed by the godlike authors of such a fictional world that their characters have been taken ill, arrested, or even killed).

Role-Playing in an Authored World

The kinds of virtual worlds I am imagining would combine the immersive pull of an authored story, like an ongoing television serial, with the open-ended agency of the MUDs. They would relieve the interactors of the responsibility for inventing the fictional world on their own.

Multiuser worlds without such external authoring run into trouble in establishing the boundaries of the illusion. For instance, one of the first experiments with a graphics-based virtual world, called Habitat, found itself immediately divided between interactors who wanted to shoot and kill one another and those who wanted to form a shared community. The organizers of the project negotiated a compromise by creating a wilderness land in which violence was routine and a town where it was outlawed. Soon the members of the town founded a church and elected a sheriff—they had essentially re-created the popular fantasy of the pioneer American West—and they immediately began quarreling over whether the townspeople could outlaw violence altogether.

The role of a central author (or team of authors) in such an environment might be to negotiate such boundary issues, for example, by insisting that the improvisational elements remain consistent with a general story line. This need not mean censoring the interactors' imagination. Moreover, the author should be able to improvise along with the interactors and to take advantage of spontaneous actions to create dramatic events appropriate to the fictional world. For instance, the Habitat world was disrupted whenever a player gained possession of the villain Death's deadly virtual gun, designed for use by the wizards of the system only. One wizard handled this situation by threatening to throw the player off the system unless he returned the gun, but a more imaginative genie faced with the same situation staged an elaborate ransoming ritual, which became a spectacle for the whole community.[7]

If participatory environments merge with authored environments, as I think they will, tensions between the author and the participants may increase. There will always be a trade-off between a world that is more given (more authored from the outside and therefore imbued with the magic of externalized fantasy) and a world that is more improvised (and therefore closer to individual fantasies). The area of immersive enchantment lies in the overlap between these two domains. If the borders are constantly under negotiation, they will be too porous to sustain the immersive trance.

A cyberdrama that combines a strong central story with active role-playing would need clear conventions to separate the area in which the interactors are free to invent their own actions from areas over which they cannot expect to have control. For instance, suppose I were to play a role in a cyberdrama based on a broadcast television series like *Babylon 5*; let us call it *Jerusalem 6*. This world would include a regular central story, like an hour-long television show (but not necessarily delivered in a single consecutive hour), and a virtual place, accessible over the Internet, in the form of a three-dimensional model of the space station. Now suppose that I decide, after watching some segments of the series in ordinary broadcast television format, to take a role in the story. I get to pick which of the alien races I want to belong to (the S'kri), what my cartoon-figure avatar looks like (cute, pointy horns), my name (Duncour), and occupation (spy).

The cyberdrama program then selects the other aspects of Duncour's life on *Jerusalem 6*, such as where I work, who the members of my family are, and which of the potential plots I will be involved in. I might witness a crime on my first day on the job or be faced with a moral dilemma when a member of an enemy race, the Karulls, falls sick at my doorstep. Since I know what is happening in the main story, I will be alert to events or documents that might affect it. Perhaps I will come across evidence of corruption in one of the most trusted members of the space station, who has been opposing a treaty with the S'kri. I would know what to do about this discovery from having seen the actors on the series deal with similar situations.

The authors of the series would have to respond to the participants' inventions while still maintaining the general outlines of the story. The participation of thousands, or perhaps even millions, of interactors in a centrally controlled story world would only be possible by limiting the kinds of roles they could play and the kinds of actions they could take. A factionalized plot, like the intricate, diplomatic maneuvering on *Babylon 5* (or in a historical romance set in the French Revolution or in a *Masterpiece Theater* saga of the English upper class), would be ideal for such a world because it would allow for many individual intrigues that fit into a larger conflict. Let us say that in our *Jerusalem* 6 story the S'Kri are fighting for their liberation from the Karulls. My character, Duncour, belongs to an underground cell (there might be thousands of such groups) allied with the dashing Grand Toff, who is one of the central characters of the series. If the Grand Toff is captured by the Karulls (as arranged by the authors), the whole cell (perhaps twenty people) would work to free him. Unlike a videogame, a role-playing world should allow each interactor to choose from several ways to go about the task, including bartering as well as fighting. The success or failure of the many small cells of interactors would influence the way this plot is developed and how long it takes to free the Grand Toff. My actions would therefore be meaningful in my own story and in the story of my small role-playing group, as well as in the central plot.

It will probably take a large, hierarchically organized writing team, comparable to the groups of writers who work on daytime television series, to generate enough plot material to maintain the interest of participants and to make sure that events in one part of the story do not anticipate or obstruct events in another. It will also take carefully ritualized patterns of participation, so that the interactors know what to expect of one another and of the authors controlling the virtual world. Most challengingly, it will take programmed occurrences that make the story world eventful and unpredictable for all of the interactors without limiting their freedom or intruding on their improvisational pleasure.

Highly ritualized interactions can actually increase the participants' freedom, rather than limiting it, by offering them more choices of coherent action. The commedia dell'arte could put on unscripted plays because the actors knew so many formulas to plug into them. Similarly, live-action role-playing games work best when experienced players guide the action into standard channels of treasure hunting, combat, social behavior, and negotiation. Among the best guides to behavior for less experienced role-players are the "mechanics" discussed in chapters 4 and 5, the formulas of behavior that provide symbolic alternatives for actions that cannot be acted out directly without disturbing the boundaries of the illusion.

For instance, suppose that when I signed on to *Jerusalem 6*, I received a character sheet that included the information that members of the Karull race found S'Kri with shiny horns like mine totally irresistible. Now let us say that I have met, in a virtual café, a Karull commander who has information on where the Grand Toff is being held. My character sheet tells me that I can seduce the commander if I spend fifteen minutes talking to him alone. I decide to help free my imprisoned leader by using my powers of seduction, and as I talk to the commander I discover that my avatar is looking more and more appealing (perhaps her delicate horns are twinkling at her companion) and that romantic music is swelling around us. No matter what we say to one another, the seduction is in progress. After fifteen minutes, the scene changes. We find our avatars have been moved to another virtual space—the back room of the bar. The lovemaking need not be shown (it is a family drama), but the interactors could role-play whatever parts of the encounter would affect the plot or be compelling for the development of character.

Since this is an adventure story, the authors would have programmed seduction sequences to include an opportunity for deception of various kinds. Let us say that in this case the system fades out on the bedroom scene and blanks the computer screen of the interactor playing the Karull commander, who understands this to mean that his character is sleeping. Meanwhile, on my own screen I see an image of his avatar sleeping. I also see that the commander's jacket has been thrown over a nearby chair and that there is something sticking out of a pocket. It would be up to me to decide if I want to risk stealing it. Perhaps it is a document telling where the Grand Toff is being held. Perhaps my companion will awaken before I can get away. Scenes like these would offer interactors the intensified immersion of collective role-playing as well as the benefit of a more surprising and eventful world than they could make up on their own.

The more structured such interactions are, the more automated such surprises could be, with the computer controlling involuntary or serendipitous events like the sleeping patterns of lovers or the opening of the pocket. A procedural author would specify the kinds of conventions (like seduction) the interactors would have access to, the conditions under which the computer would intercede and how (e.g., the computer might maximize the chance of detection when enemies are sleeping together), and perhaps the mobile viewing patterns of the interactors (thus determining which parts of the central story happen in the virtual space and who gets to see them and when). The result would be a story in which the immersive power of the given world reinforces the pleasure of the enacted role.

Of course, many people will prefer a private experience to a collective one. For them, cyberdrama would offer the opportunity to be the protagonist in a bounded world, perhaps one delivered on a digital video disc or other successor to today's CD-ROM rather than over the Net. Some worlds might be derivative from novels, films, or other media, like the Casablanca example in chapter 7, but they could also be derived from historical material, enticing Civil War buffs to live through the American 1860s or baby boomers to relive the 1950s to 1980s. Solo play would allow the interactor to explore all the stories within the limits of the world and to play all the parts until they had exhausted all the possibilities of personal imaginative engagement within a nostalgically charged situation. Although the connective and collaborative pleasures of the digital environment are the focus of much current attention, the private pleasures, like those of reading, are also likely to continue to attract us. As a domain in which we can actively participate in a responsive environment without consequence in the real world, the desktop story world may, like the novel, engage our most compelling transformational fantasies.

The Emerging Cyberdrama

I have referred to these various new kinds of narrative under the single umbrella term of *cyberdrama* because the coming digital story form (whatever we come to call it), like the novel or the movie, will encompass many different formats and styles but will essentially be a single distinctive entity. It will not be an interactive this or that, however much it may draw upon tradition, but a reinvention of storytelling itself for the new digital medium. At first the most strongly participatory cyberdrama formats may be the domain of children and adolescents, who will eagerly move from shooting games to assuming personas within dense fictional worlds, but it would be wrong to think that the format itself is merely childish. As a new generation grows up, it will take participatory form for granted and will look for ways to participate in ever more subtle and expressive stories.

Of course, the story forms described here are guesses, dependent on market forces as well as audience tastes. The term *cyberdrama* is only a placeholder for whatever is around the corner. The human urge for representation, for storytelling, and for the transformational use of the imagination is an immutable part of our makeup, and the narrative potential of the new digital medium is dazzling. As the virtual world takes on increasing expressiveness, we will slowly get used to living in a fantasy environment that now strikes us as frighteningly real. But at some point we will find ourselves looking through the medium instead of at it. Then we will no longer be interested in whether the characters we are interacting with are scripted actors, fellow improvisers, or computer-based chatterbots, nor will we continue to think about whether the place we are occupying exists as a photograph of a theatrical set or as a computer-generated graphic, or about whether it is delivered to us by radio waves or telephone wires. At that point, when the medium itself melts away into transparency, we will be lost in the make-believe and care only about the story. We will not notice it when it happens, but at that moment—even without the matter replicators—we will find ourselves at home on the holodeck.

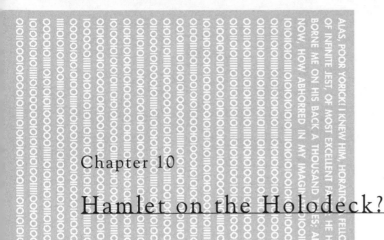

Chapter 10

Hamlet on the Holodeck?

We return to the question raised by Aldous Huxley at the moment movies began to speak: Will the stories brought to us by the new representational technologies "mean anything" in the same way that Shakespeare's plays mean something, or will they be "told by an idiot"? We have seen that the emerging cyberdrama entertainment forms need not resemble Huxley's "feelies" but could instead offer satisfactions continuous with those we receive from established narrative formats. Can we also imagine a cyberdrama that would develop beyond the pleasures of a compelling entertainment to attain the force and originality we associate with art?

We often assume that stories told in one medium are intrinsically inferior to those told in another. Shakespeare and Jane Austen were once considered to be working in less legitimate formats than those used by Aeschylus and Homer. One hundred years after its invention, film art still occupies a marginal place in academic circles. The very activity of watching television is routinely dismissed as inherently inferior to the activity of reading, regardless of content.[1] But narrative beauty is independent of medium. Oral tales, pictorial stories, plays, novels, movies, and television shows can all range from the lame and sensationalist to the heartbreaking and illuminating. We need every available form of expression and all the new ones we can muster to help us understand who we are and what we are doing here.

The real literary hierarchy is not of medium but of meaning. We focus so inappropriately on the worth of the various media in part because the last quarter of the twentieth century has brought a general crisis of meaning. As Toni Morrison has so aptly put it, we are living in a time when "to mean anything is not in vogue."[2] Commercial forces favor simplistic stories over more authentic engagement with the world. Academic theorists reduce literature to a system of arbitrary symbols that do not point to anything but other texts. But in our ordinary lives, we do not experience the world as a succession of signifiers any more than we experience it as a succession of car chases. In our ordinary lives, we turn to stories of every kind, again and again, to reflect our desires and sorrows with the heightened clarity of the imagination. We will bring these same expectations to digital narrative.

In trying to imagine Hamlet on the holodeck, then, I am not asking if it is possible to translate a particular Shakespeare play into another format. I am asking if we can hope to capture in cyberdrama something as true to the human condition, and as beautifully expressed, as the life that Shakespeare captured on the Elizabethan stage.

Procedural Authority

The most important element the new medium adds to our repertoire of representational powers is its procedural nature, its ability to capture experience as systems of interrelated actions. We are now engaged in establishing the building blocks of a procedural medium, the musical figures that may someday grow into a symphonic form. We are learning how to create characters by modeling their behaviors, how to create plots by establishing the rules by which things should happen, and how to structure the participation of the interactor into a repertoire of expressive gestures.

The notion of a procedural medium that provides the satisfactions of art takes some getting used to. We will need time to grow accustomed to combining participation with immersion, agency with story, and to perceiving the patterns in a kaleidoscopic fictional world. Most of all, the procedural medium will challenge our notions of authorship. In a print model, we think of an authored environment as fixed and not open to variation. A mutable, kaleidoscopic world can feel to some like an unauthored world.

We like to know that there is a ruling power in control of an imaginary universe, and it makes us uncomfortable if the author seems to abdicate the role. Some experiments with nonlinear form in linear media have violated this expectation. When the movie *Clue* was released in multiple versions, each with a different solution to the murder, viewers felt cheated rather than intrigued. When Italo Calvino mutates his plot and characters with every chapter of *If on a Winter's Night a Traveler,* my students lose interest in the story. People who get great pleasure from books and film are often hostile to the very idea of digital narrative because they expect it to be disappointing in just this way.

Yet once we understand simulations as interpretations of the world, the hand behind the multiform plot will feel as firmly present as the hand of the traditional author. With familiarity we will come to realize that the procedural author can shape a juxtaposition or a branch point in a multiform story as artfully as a traditional author shapes a speech in a play or a chapter in a novel. Already in the gaming world there are clear auteurs, creators with characteristic and original style as well as strong technical mastery. To play *Mario Brothers* or *King's Quest* or *Myst* is to open ourselves to the vision of the shaping author in the same way we open ourselves to the author's voice in a novel.[3] Just as we have only recently learned to think of the solitary reader as playing an active role through imaginative engagement with the story, so too are we just beginning to understand that the interactor in digital environments can be the recipient of an externally authored world.

A George Eliot or Leo Tolstoy or William Shakespeare of the future could create kaleidoscopic worlds of dazzling variety that will display the coherence and unified vision we associate with great fiction. Cyberdramatists will exercise authorial control through the techniques of procedural authorship (described in chapters 7 and 8), which would let them dictate not just the words and images of the story but the rules by which the words and images would appear. But would we feel silly playing a role within an environment created by a great artist? Will we feel called upon to act like Olivier or to produce Bard-like lines? Not necessarily. Future audiences will take it for granted that they will experience a procedural author's vision by acting within the immersive world and by manipulating the materials the author has provided them rather than by only reading or viewing them. They will welcome the choice-points in the narrative as dramatically heightened moments shaped for them with the same artistry that we now expect in the editing of a film. They will accept their exercise of agency as part of the aesthetic experience in the same way that we now take it for granted that we have to walk around a Degas sculpture to experience its full beauty rather than merely stand in front of it as we do with his paintings.

Traditions of Virtuosity

The model of authorship I have just elaborated is that of the single genius, the hero-writer celebrated by the Romantic poets, a model we have come to associate most strongly with the figure of Shakespeare, just as we have come to associate him with the noblest achievements of culture, achievements many feel to be threatened with oblivion by the "brave new world" of technology. We forget that Shakespeare did not write books; he wrote plays and spent his life in the collaborative medium of the theater, shaping his characters to fit the strengths of his acting company.

The Shakespearean stage, with its thrillingly scripted text and single author, is often contrasted with the other great theatrical achievement of the time, the Italian commedia dell'arte, which derived its power from the improvisational performance skills of its actors. In a high culture model of art, what gets preserved is a particular story in a fixed form, to be repeated verbatim. In a folk model, like the bardic tradition discussed in chapter 8, it is the forms that get passed down, to be altered and repurposed by each succeeding generation.

Shakespeare's plays are still performed and read. Although they reflect the formulaic theater of his time and the acting strengths of particular members of his company, and although they exist in variant versions and are full of appropriated material, we recognize them as expressing the vision of a single virtuoso mind. The commedia dell'arte plays are now rarely performed, but the theatrical traditions of the commedia can be seen in Mozart's *Figaro* or in a Marx brothers skit or in the improvisational moviemaking of Robert Altman or Mike Leigh. Its artistry took the form not of the single controlling genius but of virtuoso collaboration.[4]

Cyberdrama presents us with the possibility for both kinds of virtuoso authorship and for many mixtures of the two. Today's role-playing environments (electronic and live action) are like the commedia in that they base their dramatic improvisations on written materials. For though the commedia was a folk form, it was very much a product of the age of print in that it was performed by highly literate actors who had memorized passages of poetry and prose, which they repeated verbatim or used as models for improvised creation. Like the role-playing interactors in the MUDs, the commedia dell'arte actors used their literary models as quarries from which to draw material that helped them elaborate their characters.

Shakespeare is our touchstone for literary art because his stories move us across centuries and even across cultures. But a tradition of narrative art is also fed by stories that enjoy a much more circumscribed audience, such as a small circle of friends or family members. The Brontë children began by making up their imaginary kingdoms for one another. Without Angria and Gondal there would have been no Thornfield or Wuthering Heights. Orally transmitted stories often have a stronger resonance for members of their originating culture than they can ever have for outsiders. The stories that people make up collaboratively in virtual environments are of this tribal nature; they may seem trite or derivative to an outsider, but they can be riveting and emotionally resonant for the participants.

Like other folk traditions, the role-playing tradition aims for ephemeral performance that preserves not particular scenes but the conventions of interaction. Over the next century these conventions may rapidly evolve into a more expressive repertoire of improvisational structures. We may come to think of cyberdrama in all its variations as an essentially collaborative art form. Perhaps a group of role-players will be like a commedia dell'arte troupe, more skillful in combination than any of them would be alone. Perhaps such groups will coalesce around a few star performers whose invention and dramatic force provide creative direction for the less skillful interactors. Perhaps, in time, role-playing might experience a Homeric transition: a consolidation of a collectively improvised tradition into a single repeatable work. But whether or not this happens, the on-line role-playing contributions of amateur improvisers will lead to new formulas of interactions that will feed into the general expressiveness of the medium.

Formulaic Invention and Originality

In thinking about both narrative traditions and computational environments, we have necessarily been thinking in terms of formulas. The creation and refinement of narrative formulas are the necessary preconditions for the creation of any great work of art. Without those repetitive revenge plays that are now only read in graduate school, we would not have *Hamlet*. Nor could *Jane Eyre*'s psychological realism exist without the simplistic gothic tradition of menaced heroines locked up in spooky castles. Literary stereotypes are like rough sketches of the world, which the next generation or the more capable artist can modify and elaborate. The eighteenth century, confronting the demons of the psyche for the first time in a postreligious frame-work, created salacious stories of blaspheming villains imprisoning virgins; the nineteenth century drew on these story elements to ex-press the courage of the nascent feminists. Entertainment forms try to give energy and novelty to stereotypical formulas, but art reshapes the formulaic to conform more closely to the world of experience. Yet these activities are closely intertwined and dependent upon one an-other. We could not have the breakthrough achievement of a work of lasting art without the originality and inventiveness of less ambitious stories. Formulaic entertainment and form-shattering art are both embedded in a cultural repertoire of story patterns. Electronic narra-tive will only translate that repertoire into a new arena.

Charlotte Brontë's adolescent stories were powerfully moving to her and her brother. The characters of Angria were so strong for her that she sometimes experienced them as a hallucination. But if she had died before she wrote *Jane Eyre,* her visions would not have passed into the general culture. The difference between the Brontë juvenilia (discussed in chapter 6), and *Jane Eyre* is the difference be-tween derivative and rigidly formulaic expression and an original work of transforming genius. When Caroline Vernon carries her mother upstairs, it is a powerful expression of Brontë's own conflicts and a sharply rendered fantasy. When Jane Eyre cries out, "I care for myself!" and rushes from Rochester's mansion, it is a powerful expres-sion of Victorian social realities and of the enduring human conflict between passion and autonomy. Charlotte Brontë could not have written *Jane Eyre* solely within the formulas of her juvenilia any more than she could have written it solely within the form of the courtship novel as Jane Austen had perfected it. Her work survives because it transcended both kinds of restrictions: the limitations of privatized, formulaic expression and the limitations of a preexisting exemplary narrative format.

We may be at the juvenilia stage of electronic narrative for some time yet, as we gain practice in procedural virtuosity. But we have come far enough in establishing the traditions that will nourish future storytellers that we can begin to speculate on what the medium might someday offer us in a more realized art.

New Beauty, New Truth

Every age seeks out the appropriate medium in which to confront the unanswerable questions of human existence. We cannot limit ourselves to Elizabethan or Victorian forms any more than Shakespeare could have written within the conventions of the Aristotelian tragedy or the medieval passion play. When Hamlet stands alone on the stage, pondering whether "to be or not to be," he personifies, among other things, the Renaissance fascination with thinking itself and with the separateness of the individual life. Shakespeare's extensive use of soliloquy in *Hamlet* is an appropriate technical innovation to capture this newly experienced solitude. Soliloquy on the Elizabethan stage was used mostly for letting the villain or revenge seeker tell the audience what he was up to, but in *Hamlet* Shakespeare uses it to make what his protagonist is thinking more dramatic than what he is or is not doing. Although Hamlet is not the first character to reveal his thoughts on stage or to utter a soliloquy, his particular expression of meditative self-consciousness is both original and universal. It represents a truth about human experience that could not be told before.[5]

Hamlet's soliloquies, like Jane Eyre's rush away from Thornfield, epitomize what a new narrative format can offer us. What similarly revealing construction of the world might we expect from a fully realized cyberdrama? What aspects of our inner and outer lives await the expression of a future cyberbard?

The most ambitious promise of the new narrative medium is its potential for telling stories about whole systems. The format that most fully exploits the properties of digital environments is not the hypertext or the fighting game but the simulation: the virtual world full of interrelated entities, a world we can enter, manipulate, and observe in process. We might therefore expect the virtuosos of cyberdrama to create simulated environments that capture behavioral patterns and patterns of interrelationships with a new clarity. The tragic story of the suicide described in chapter 6 suggests the kinds of subject matter that might be appropriate for expression as a complex system: the thought patterns of a particular mind, the web of family relationships. But perhaps the new medium can take us even further in both directions, looking deeper into the human mind and encompassing even more of the external social world.

One major trend in literary history from the time of Shakespeare onward can be imagined as a camera tracking in from a medium shot to an extreme close-up on human consciousness itself. After reading the wildly digressive monologue of Sterne's Tristram Shandy or the exquisite moral discriminations of a Henry James heroine or the richly textured stream of consciousness captured by Virginia Woolf, it is hard to believe that we could penetrate any further into the workings of the mind. But twentieth-century science has challenged our image of ourselves and has perhaps outrun our ability to imagine our inner life. A linear medium cannot represent the simultaneity of processing that goes on in the brain—the mixture of language and image, the intimation of diverging possibilities that we experience as free will. It cannot capture the secrets of organization by which the inanimate somehow comes to life, by which the neural passageway becomes the thought.

Perhaps a great procedural virtuoso of the next century will be able to bring these elusive patterns of the mind into sharper focus. A stream-of-consciousness cyberdrama (like the exploration of Rob's mind, described in chapter 6) could perhaps center on the miracle of conversion, on how we can sometimes shift our perceptions of the world from a momentary revelation, or on how we manage to transform ourselves by redescribing who we are at just the right level of self-awareness.[6] Perhaps a James Joyce of the electronic age will find a way to take us so deeply into a single particular consciousness that we will be able to trace the exact border between what we think of as brain and what we think of as mind. Perhaps a future

Dickens will create a set of chatterbots so comically expressive of the mechanical nature of human thought and yet so endearing that the humblest circuits of the amygdala will come to hold a whimsical charm for us. We fear the computer as a distorting fun house mirror of the human brain, but with the help of the narrative imagination it might become a cathedral in which to celebrate human consciousness as a function of our neurology.

The narrative imagination has the power to play leapfrog with analytical modes of understanding. Ancient myths described the power of the sun god before we understood photosynthesis or the physics of light. Shakespeare created *Hamlet* without benefit of Freud. In the same way, the coming cyberdrama may help us reconcile our subjective experience of ourselves with our rapidly expanding scientific knowledge of biology. It may come up with the metaphors of process that will restore the sense of human individuality to our model of the mind. A computer-based literature might help us recognize ourselves in the machine without a sense of degradation.

The kaleidoscopic powers the computer offers us, the ability to see multiple patterns in the same elements, might also lead to compelling narratives that capture our new situation as citizens of a global community. The media explosion of the past one hundred years has brought us face-to-face with particular individuals around the world without telling us how we are to connect with them. The exploration of space has taught us that we are all part of a single society but not how to find our place in it. The capaciousness and specificity of the computer offers us a way to model the behavior of single individuals within great groups of people, to make up fictional worlds in which we can enact the confusions of membership in a newly visible yet overwhelmingly various worldwide humanity.

D. H. Lawrence argued that "the novel is the highest example of subtle inter-relatedness that man has discovered. Everything is true in its own time, place, circumstances, and untrue out of its own time, place, circumstances."[7] The novel can put things in their place, can let us figure out what is right and wrong by offering us a specific context for human behaviors. But in a global society we have outgrown our ability to contextualize. We are tormented by our sense of multiple conflicting frameworks for every action. We need a kaleidoscopic medium to sort things out.

Not only is the computer the most capacious medium ever invented, but it also allows us to move around the narrative world, shifting from one perspective to another at our own initiative. Perhaps this ability to shift perspectives will lead to the technical innovation that will rival the Shakespearean soliloquy. Cyberdramatists of the future could present us with a complex world of many characters (like a global Victorian novel) and allow us to change positions at any moment in order to see the same event from the viewpoint of another character. Or they could let us enter a particular town over and over again in the guise of many different individuals, enabling us to see how differently the same people present themselves to us. We might be given a compelling role within the environment that confers upon us the ability to fluidly switch between viewing the world through our own character's eyes and viewing our character through the eyes of others. Or perhaps a cyberdramatist of the future will find a way to show us not just the large battlefield and the single soldier (as Tolstoy does in *War and Peace*) but also the processes by which large historic events emerge as the sum of many much smaller causes (as Tolstoy strove to convey in his interpolated essays but could not dramatically capture). All of these story patterns would be ways of enacting the contemporary human struggle to both affirm and transcend our own limited point of view.

Finally, the experience of the Habitat community described in chapter 9 suggests that the collective virtuosity of the role-playing worlds may provide a tradition of stories around the themes of violence and community. The violent gaming culture that now characterizes much of cyberspace is likely to spread as the Internet gains speed and bandwidth. Teams of combatants from every corner of the globe will blast each other's avatars with ever more macho digital weapons; the narrative formulas of combat tied to disturbingly lurid images will continue to proliferate. At the same time, the communal aspects of cyberspace are also growing rapidly, with people eager to construct utopian fantasy worlds that they can share with one another. The Internet is therefore likely to serve as a global stage for conflicts between these two groups, turning the struggle between the blasters and the builders into a kind of worldwide morality play.

There are probably not two more difficult things to predict in this world than the future of art and the future of software. These visions of the future can only be speculations, extrapolations from the current environment, which is shifting even as I write. The computer is chameleonic. It can be seen as a theater, a town hall, an unraveling book, an animated wonderland, a sports arena, and even a potential life form. But it is first and foremost a representational medium, a means for modeling the world that adds its own potent properties to the traditional media it has assimilated so quickly. As the most powerful representational medium yet invented, it should be put to the highest tasks of society. Whether or not we will one day be rewarded with the arrival of the cyberbard, we should hasten to place this new compositional tool as firmly as possible in the hands of the storytellers.

Notes

Introduction

1. Lawrence, "Why the Novel Matters," 105.
2. For instance, I wrote about the works of the feminist Victorian novelist George Meredith and edited reprint editions of *The Englishwoman's Review,* the Victorian feminist magazine of record, and of *Miss Miles,* a feminist novel written by Charlotte Brontë's closest friend, Mary Taylor.
3. Murray, *Strong-Minded Women.*
4. I was particularly influenced by the work of Claire Kramsch, who pioneered communicative language learning methods, and Peter Elbow and Linda Flowers, both of whom pioneered the teaching of writing as a process-centered, rather than product-centered, activity.
5. The programming language was LISP (LISt Processing language), a language designed in the 1950s by John McCarthy for use in artificial intelligence research. The introductory software engineering course at MIT (6.001) uses a dialect of LISP to train students in designing software systems. The instructors have been known to wear wizards' hats and to display yin/yang signs to describe the almost magical interpenetration of data and procedures in LISP. See Abelson and Sussman, *Structure and Interpretation of Computer Programs.*
6. LOGO, which is based on LISP, allows children to master procedural thinking and abstract concepts like recursion—the ability of a function to call itself, which LISP allows—in the course of building things. In addition to acquiring mathematical concepts by choreographing on-screen sprites, children learned principles of geometry by writing computer scripts for a turtle that drew lines as it moved across the floor. In his later work, Papert established a partnership with the Lego company and created systems that let children build and script their own robots. See Papert, *Mindstorms,* for Papert's early microworld theory and practice, and Turkle, *The Second Self*

(especially pp. 141–54), for descriptions of children working with LOGO microworlds.

7. Lippman, "Movie-Maps."

8. The Athena Language Learning Project 1983–96 was funded by the Annenberg/CPB Project, the National Endowment for the Humanities, Apple Computer, and the Consortium for Language Teaching and Learning. It explored three technologies for language learning: natural language processing, speech processing, and interactive video. Natural language processing proved impractical for student use, speech processing proved promising for adult learner pronunciation practice, and interactive video was widely successful with language teachers and learners; see Murray, "Lessons Learned." The most successful product of this effort to date is *A la rencontre de Philippe* (available on videodisc), which has won numerous awards, including a Gold CINDY and an Educom Special Recognition Award, and was codesigned by Gilberte Furstenberg, Ayshe Farman-Farmaian, Stuart Malone, and me. It is an interactive narrative with seven possible endings and many possible paths. For a description of *No recuerdo,* a more complex ALLP narrative currently being redesigned for CD-ROM, see Morgenstern and Murray, "Tracking the Missing Biologist."

9. The MIT Shakespeare Electronic Archive Project is funded by the National Endowment for the Humanities and the Andrew W. Mellon Foundation and directed by Peter Donaldson, Larry Friedlander, and me. It is a successor to Larry Friedlander's Stanford Shakespeare Project of the 1980s, a videodisc project that linked multiple performances of key scenes from the plays with the text. The MIT Shakespeare Electronic Archive Project is linking modern editions, photofacsimiles of early editions, art collections documenting performances, and film performances. For a description of the project, see Donaldson, "The Shakespeare Interactive Archive."

The Virtual Screen Room project, funded by the National Endowment for the Humanities and directed by Henry Jenkins, involves the development of a prototype of a multimedia learning environment that would replace the conventional textbook for introducing students to techniques and critical concepts in the study of film art.

Among the many impressive projects created elsewhere that have confirmed my sense of the usefulness of the medium for teaching things that could not be as well conveyed by other means are Gregory Crane's *Perseus Archive of Ancient Greece,* developed at Harvard and Tufts; James Noblitt's System D, for learning French, developed at Cornell University and at IBM's Institute for Academic Technology, affiliated with the University of North Carolina at Chapel Hill; and George Landow's Dickens Web and In Memoriam Web, developed as part of the Hypermedia Project at Brown University.

10. The variorum Shakespeare is overseen by a committee of the Modern Language Association. The identity of my hosts has been disguised, and this

momentary reaction does not reflect the current attitudes of the committee. The anecdote captures the anxiety felt in this and similar circles as electronic publication began to be taken seriously in the early 1990s.

Chapter 1. Lord Burleigh's Kiss

1. From the episode "Persistence of Vision" in the series *Star Trek: Voyager.* (See bibliography for production credits.)
2. *Star Trek* was created by Gene Roddenberry. The original series went on the air in 1966 and lasted for three seasons and seventy-nine episodes. As of 1997, the franchise includes seven movies and three additional television series—*Star Trek: The Next Generation* (debut, 1988), *Star Trek: Deep Space 9* (debut, 1993), and *Star Trek: Voyager* (debut, 1995). Although *The Next Generation* went off the air after seven seasons, it continues to be seen in reruns, as does the original or "classic" series. (See bibliography for production credits.)
3. Lawrence Krauss, in *The Physics of Star Trek* (pp. 99–108), has analyzed the series' combination of plausible and wildly fanciful inventions. He considers the three-dimensional images of the holodeck possible although its use of "matter replicators," which make things out of thin air, is unrealistic. We could therefore imagine a future holodeck theater in which images would surround us but would be incapable of being touched. Janeway would not be able to drink holodeck tea or sit on a holodeck parlor chair, let alone receive a holodeck embrace from Lord Burleigh.
4. Although *Star Trek* holodeck adventures were originally referred to as holodeck programs, the producers have come to distinguish between programs, which simulate a place and its inhabitants, and "holonovels," which offer complex narratives. Crew members might run a holodeck program of sailing on Lake Como or of a nineteenth-century French billiard parlor, complete with flirtatious men and easy women, to enjoy as a recreation environment, but they run a holonovel to participate in a story as Beowulf or as a Victorian governess.
5. The Lucy Davenport story appeared in three episodes of *Star Trek: Voyager*: "Cathexis," "Learning Curve," and "Persistence of Vision." (See bibliography for production credits.)
6. *The Republic of Plato,* chapter XXXVI.
7. Denby, "Buried Alive: Our Children and the Avalanche of Crud," 48–58.

Chapter 2. Harbingers of the Holodeck

1. McLuhan compares the incunabula to the "horseless carriage," because they were often attempts to reproduce the manuscript by mechanical means rather than to invent a new form (see McLuhan, *The Gutenberg Galaxy,* 153). It is also true that the handwritten manuscript book had already evolved some of the essential elements we associate with printed books, such as regularity of presentation, separate pages, and large letters to indicate divisions in topics. If the incunabula represents the book in

swaddling clothes, the manuscript book represents its embryonic state, and both are very recognizable forerunners of the more mature format. See Eisenstein, *The Printing Revolution,* for a scrupulous examination of the continuities and discontinuities in the transition to print (especially pp. 3–40) and Bolter (especially pp. 63–74) for a consideration of the similarities and differences between the papyrus scroll, the manuscript codex, and the printed book.

2. Foreword by Irving Howe in Schwartz, *In Dreams Begin Responsibilities,* viii.
3. Danny Rubin and Hal Ramis, the writer and director of *Groundhog Day,* had to fight with the studio to avoid having to put a "gypsy curse" explanation of Phil's predicament into the movie (see Lippy, "A Talk with Danny Rubin," 183, and Lippy, "Harold Ramis on Groundhog Day," 53). Rubin's premise arose from his thinking about changing the rules of life for "the type of guy . . . who just can't seem to get past his adolescence." He wondered, "If someone like that really lived long enough, would he ever get over it? . . . At what point would he get bored with himself and try to become something else?" (Lippy, "A Talk with Danny Rubin," 183).
4. Jenkins, *Textual Poachers.* See especially chapter 6, "Welcome to Bisexuality, Captain Kirk."
5. For a description of the role-playing game culture of the 1970s and early 1980s, see Fine, *Shared Fantasy.*
6. Between June 1996 and February 1997, Boston dinner theater audiences were offered the role of wedding guest at two different "comedy weddings," bachelorette party guest at a third event, mourner at a comedy wake, juror at a murder trial, and detective/cruise-goer on a mystery ship. Outside of dinner theater, audiences have been addressed as schoolchildren, in Christopher Durang's *Sister Mary Ignatius Explains It All to You,* and as PTA members, in A. R. Gurney's *The Rape of Bunny Stuntz.* In *Tamara,* which was produced off-Broadway in the 1980s, audience members had to follow the actors around a town house.
7. For a psychosocial analysis of role-playing in MUDs, including a comparison with live-action role-playing, see Turkle, *Life on the Screen* (especially pp. 177–209). Turkle sees MUDs as a psychological workshop for trying on multiple identities. Rheingold, *Virtual Community,* covers the issues of alternate identities from a journalistic perspective (pp. 145–75); his general description of the Internet as a "virtual community" is very relevant to the way in which many people experience the MUDs.
8. Interview on "Roller Coaster," *Nova* PBS series, 1996.
9. George Landau suggested the useful term *lexias,* taking it from Roland Barthes, who invented it as a term for "reading unit" as part of his theory of texts. See Landau, *Hypertext,* 4, 52–53, and Barthes, *S/Z,* 13.
10. Zakarin, "The Spot."
11. See Harpold, "Conclusions," and Douglas, "How Do I Stop This Thing?" for appreciative readings of *Afternoon.*
12. Diamond Park is the creation of the Mitsubishi Electric Research Labora-

tory in Cambridge, Massachusetts. For more information see http://www.merl.com/projects/dp.

13. See Laurel, Strickland, and Tow, "Placeholder: Landscape and Narrative in Virtual Environments" or see http://web.interval.com/projects/placeholder/.

14. See Laurel, *Computers as Theater* for a compelling statement of her aesthetics.

15. ALIVE stands for Artificial Life Interactive Video Environment, a project intended "to demonstrate that virtual environments can offer a more 'emotional' and evocative experience by allowing the participant to interact with animated characters" in a virtual world that the interactor enters without having to wear any equipment. See Maes, "Artificial Life Meets Entertainment."

16. The characters are built on models based on animal behavior. The architecture of the characters is discussed further in chapter 8. See Blumberg, "Action-Selection in Hamsterdam," and Blumberg, "Old Tricks, New Dogs."

17. For the work of the Oz group, see Bates, "The Role of Emotion in Believable Agents"; Bates, "Virtual Reality, Art, and Entertainment"; and Kelso, Weyhrauch, and Bates, "Dramatic Presence."

18. Weyhrauch, "Guiding Interactive Drama."

Chapter 3. From Additive to Expressive Form

1. The French title of the film is *L'Arrivée d'un Train à la Ciotat.* The film is discussed in most film textbooks (e.g., Mast and Kawin, *A Short History of the Movies,* 22, and Cook, *A History of Narrative Film,* 11), with the audience sometimes described as "ducking" and sometimes "stampeding." Thomas Gunning, "An Aesthetic of Astonishment," considered the historical evidence about the reception of the film and concluded that audiences did not take the train's arrival as a real event but as an "attraction," a new and delightful kind of theatrical illusion.

2. Weizenbaum, "ELIZA, 36."

3. The researcher was Daniel G. Bobrow who was working at the Cambridge research firm of Bolt Beranek and Newman (BBN) at the time. Bobrow's account of the incident was published in the December 1968 issue of the *SIGART Newsletter* (from the Special Interests Group in Art of the Association for Computing Machinery) and is reprinted in McCorduck, *Machines Who Think,* 225, and without attribution in Boden, *Artificial Intelligence and Natural Man.* For many years the story was told at MIT in computer science courses as if it had happened to Weizenbaum himself.

4. Weizenbaum, *Computer Power and Human Reason,* 189.

5. In *Aspects of the Novel* (p. 42), E. M. Forster uses Mrs. Micawber as an example of a purely flat fictional character who will never grow beyond her single-sentence self-description. His distinction between flat and round characters is discussed in chapter 8.

6. This is an unedited extract from one of my own conversations with a PC-based Eliza, originally published in Murray, "Anatomy of a New Medium."

7. Sherry Turkle (*Life on the Screen*, 120–23) reports on an interactor who found benefit in an automated therapist called DEPRESSION 2.0, although he did not mistake the program for an actual person. The "chatterbot" Julia, however, discussed in chapter 8, has often been mistaken for a human being.

8. *Zork* was not the first adventure game for the computer. That honor goes to *Adventure*. Computer puzzle gaming began in 1972 when William Crowther, using the Fortran programming language, plotted out a cave he had explored. In 1976 Don Woods, a researcher at Stanford's Artificial Intelligence Laboratory, expanded Crowther's cave game with fictional elements drawn from Tolkien. This version was still in Fortran but was soon translated into C and installed on Unix systems in research labs and universities all around the country. *Adventure* established the basic format of a treasure hunt in which the interactor moves around a virtual space (Colossal Cave, in the original) and fights off attackers by typing in commands and receiving a description of events in return. Zork was an enthusiastic response to the excitement of *Adventure*. See Lebling, Blank, and Anderson, "Zork: A Computerized Fantasy Simulation Game," 51–59, for a description of how *Zork* improved upon the *Adventure* model by exploiting the strengths of MDL, a descendant of LISP. See Niesz and Holland, "Interactive Fiction," for the first article considering adventure games as a literary genre. See Pinsky for a description of Mindwheel, the poet's literarily ambitious use of the *Adventure/Zork* framework to create a fantasy world.

9. *Crime Story*, Quest Interactive Media; available at http://www.quest.net/crime/updated February 2, 1997.

10. *SimCity* has been faulted for, among other things, its bias against mixed-use development and its systematic denial of racial conflict. See Turkle, *Life on the Screen*, 70–73, for a good discussion of the perils of hidden assumptions in simulations in general and in *SimCity* in particular.

11. See Gary Woolf's article on Nelson and *Xanadu* in *Wired*, p. 140. Nelson takes medication for attention deficit disorder, but he rejects the name of the condition as an invention of the "regularity chauvinists" and prefers the term "hummingbird mind" for his own experience. The Woolf article describes the visionary insight and frustrating development process behind Nelson's lifelong pursuit of the perfect hypertext system.

12. Nelson, *Literary Machines*, 93.1, page 6/6.

13. The *Game of Life* was invented by the mathematician James Conway in the late 1960s and moved to the computer in the late 1970s through the efforts of Edward Fredkin of MIT. For a consideration of the science of artificial life and the philosophical questions it poses, see Emmeche, *The Garden in the Machine*. For a similar survey from a psychological standpoint, see Turkle, *Life on the Screen*, 150–58. Turkle describes the wonder she felt the first time she saw the computer animation based on Conway's rules: "I

stood alone at the screen, watched the Game of Life, and felt like a little girl at the ocean's edge" (p. 155).

14. T. S. Eliot uses the term *objective correlative* in a 1919 essay on Hamlet. He considers the play a failure because Shakespeare did not find an adequate objective correlative for the emotions it contains. Hamlet's emotions are "in excess of the facts" and therefore not well communicated to us. For Eliot a successful objective correlative is "a set of objects, a situation, a chain of events which shall be the formula of that particular emotion; such that when the external facts . . . are given, the emotion is immediately evoked " (Eliot, ed., *Hamlet*, 48). The pieces of the work of art fit together in order to express the otherwise inexpressible, to transfer the experience of an emotion from writer to audience.

Chapter 4. Immersion

1. Silent reading itself was not new in the Renaissance (as McLuhan wrongly assumed), but it was "increasingly more pervasive and ever more elaborately institutionalized after the shift from script to print" (Eisenstein, *The Printing Revolution*, 91–92). The notion that Don Quixote's brain would dry up from prolonged reading and that he could lose touch with reality reflects the fear that this solitary and internal behavior provoked. Just as television and now the computer raise concerns in our day, so too was there concern in the Renaissance over the vast quantities of material that were suddenly available to people who did not have access to them before. The early printers eagerly turned to the materials of the Middle Ages to supply what electronic publishers now refer to as "content" for the capacious new medium, hence the availability of the Romances that derange Don Quixote. He represents a peculiarly modern fear: that exposure to vast amounts of fictional material (enough so that he can read night and day for months at a time) in an activity that isolates him from other people will result in his substituting the illusory world for the world around him. For McLuhan's commentary on the change in attitudes brought about by silent reading rather than reading aloud in groups, see *The Gutenberg Galaxy*, 84–90.

2. I am not saying that the juror was delusional. In showing up in Star Trek uniform for jury duty at a trial involving a political scandal, she was pointing to a world with a higher ethical standard. Like Don Quixote's (and more explicitly in her case), her allegiance to a fictional reality was a form of resistance and critique of the values of those around her. See Jenkins, "The Politics of Fandom," 15.

3. See Turkle, *The Second Self*, for an analysis of the computer as an "evocative object" and her *Life on the Screen* for a study of the ways in which MUDders and newsgroup members assume alternate identities, fall in love, and engage in otherwise inhibited behavior on the computer.

4. The term *liminal* is an anthropological term taken from the Latin word for

"threshold." It is used to describe mythopoetic experiences in which an object, a ritual, or a story occurs somewhere between the world of ordinary experience and the world of the sacred (see Turner, *The Ritual Process*). I am using the term to indicate the threshold between the world we think of as external and real and the thoughts in our mind that we take for fantasies. When a storyteller captures our attention and induces a deep state of absorption, we are in a threshold state, filled with real sensations and emotions for imaginary objects. This is the immersive trance.

5. Winnicott, *Playing and Reality*, 123.
6. "Better Than Life" was the title of episode 8 of the British television series *Red Dwarf*. The series was shown in the United States on public television (initial broadcast, September 13, 1988).
7. The term *hyperreality* was coined by the French postmodern theorist Jean Baudrillard (see *Simulations*) to refer to the substitution of the reproduction for the actual object, as, for instance, when someone prefers a copy of the *Mona Lisa* to the painting itself. It is a useful concept for thinking about the dizzying merger between the real and the simulated, as in events staged for the media, like a presidential inauguration; crimes committed in imitation of movies; and places like Disneyworld's Main Street, which is based on a combination of cultural fantasy and corporate merchandising rather than on social reality. Umberto Eco wrote on Disneyworld and similar venues and on the audience's response to multiple narrative clichés in cult movies like *Casablanca* as forms of hyperreality (see *Travels in Hyperreality*). Postmodern writing about the digital world often assumes that it is intrinsically hyperreal. But hyperreality is less a property of a particular medium than a way of experiencing media in general: a teetering on the border between a powerfully present illusion and a more authentic but flickeringly visible ordinary world.
8. See Morse, "Nature Morte," especially p. 215, and Dove and Mackenzie, "Archeology of a Mother Tongue." Similarly, in John McDaid's very early experimental computer narrative *Uncle Bunny's Phantom Funhouse*, one screen simulated a system crash and another looked like the script of the underlying programming code but was actually a poem. See Stuart Moulthrop's appreciative description of it in "Toward a Paradigm for Reading Hypertexts."
9. Canemaker, *Winsor McCay*, 13–15, 145–46.
10. Since the 1970s literary critics have been increasingly aware of the role of the reader in actively shaping the experience of a novel (and of the audience members in shaping the experience of a play or film). There are several overlapping strains in this analysis. Some critics focus on the reader's perception of the formulas of the narrative itself (like the marriage plot or the detective story) and emphasize the many ways in which these formulaic elements can be perceived and combined within any single work. Others focus on the emotional projections of the reader, how the individual's hopes and fears determine the shape of a particular reading. Still others

emphasize the cognitive activity of the reader in fitting the work of art into existing schemata, or frames of reference, and in trying to arrange its elements into a coherent whole. Although there is no single terminology or methodology that has emerged from this inquiry, there is a shared awareness that the act of reading (or viewing) is far from passive and requires considerable emotional and cognitive activity and a sophistication in the formulas of storytelling. See Eco, *The Role of the Reader;* Holland, *The Dynamics of Literary Response* and *Five Readers Reading;* and Iser, *The Act of Reading.*

11. The program is *A la rencontre de Philippe* (listed in the Digital Works section of the bibliography). The virtual phone had a strong reality in the mid-1980s, but if we were designing a similar system today, we would have to include the ability to record voice mail, since students now take clickable pictures for granted and would not experience it as a virtual object unless they could talk into it. The hyperreal or virtual is always a moving target. What seems eerily present today will be taken for granted tomorrow. When Sherry Turkle interviewed children in the early 1980s, she found that they thought of the computer as alive or as posing intriguing questions about what is alive and what is not (see *The Second Self,* chapter 1). By the 1990s, children the same age had a ready answer: machines cannot be alive (see *Life on the Screen,* p. 83). The culture had assimilated computers in that time as an ordinary form of representation.

12. See Frye, *Anatomy of Criticism,* 288–93, for a discussion of the spectacle (or masque) as a dramatic genre.

13. For a description of the Woggles project (also called "Edge of Intention"), see Bates, "Virtual Reality, Art, and Entertainment," and http://www.cs.cmu.edu/afs/cs.cmu.edu/project/oz/web/worlds.html#woggles.

14. Turkle, *Life on the Screen,* 189. Note also Rheingold's use of the metaphor of "homesteading on the electronic frontier" for use of the Internet, and notice the use of the term *home page* for the autobiographical postings people put on the World Wide Web.

15. Although participation in MUDs and LARPs are very similar activities, it is rare to find people who enjoy both of them. More often, MUDders express scorn at the childishness of live-action players carrying their foam rubber swords through college hallways and see themselves as engaging in a literary activity and community building. The MUDders take pride in constructing an enduring virtual world as a technical accomplishment and as a positive social activity. Live-action enthusiasts tend to look upon MUDders as unsocialized nerds; they stress the dramatic value of their own activities and the importance of looking a person in the eye as they enact their role. But the real difference is only a matter of which environment happens to create the fragile state of enchantment for a particular person. The activity itself, in both cases, ranges from an entertaining way of socializing to an intense imaginative engagement in a collaborative improvisation.

16. Winnicott, *Playing and Reality,* pp. 60–61.

17. See Turkle, *Life on the Screen*, 250–53, for a discussion of a virtual rape and its aftermath.

Chapter 5. Agency

1. Deleuze, Gilles, and Felix Guattari. *A Thousand Plateaus: Capitalism and Schizophrenia*. Minneapolis: University of Minnesota Press, 1987.
2. Moulthrop, "Containing Multitudes," 1.
3. The authoring environment Storyspace, distributed by Eastgate Systems, was designed by Jay David Bolter, John Smith, and Michael Joyce. It has the virtue of allowing the writer to see a text as a series of lexias displayed as boxes with links between them in the form of arrows. Storyspace played an important role in popularizing hypertext writing in academic circles before the arrival of the World Wide Web.
4. The robbery story is "Flow" by Robert Frederick.
5. See Thorburn, "Television Melodrama."
6. I am grateful to Matthew Gray for telling me about this puzzle in *Zork II*.
7. Of course, as the reader response critics (see Holland, *The Dynamics of Literary Response* and *Five Readers Reading;* Iser, *The Act of Reading;* and Eco, *The Role of the Reader*) remind us, reading and viewing are not passive experiences at all but require us to construct the story in an active manner. There is still a difference, however, between this emotional and cognitive activity and the external actions we take in a game or electronic narrative. Perhaps the most important difference is that in the latter we are conscious of our activity, which changes our relation to the story.
8. For more on the interpretation of games, in a more anthropological rather than literary context, see Avedon and Sutton-Smith, *The Study of Games.*
9. Tetris (Spectrum Holobyte, 1987) was originally designed by Alexey Pazhitnov in Moscow in 1985. Thanks to Elizabeth Murray and Henry Jenkins for sharing their Tetris experiences with me.
10. For a consideration of the agon as a characteristic organizing form of oral culture, see Ong, *Orality and Literacy*, 43–45.
11. James Aspnes invented TINYMUD, a software environment that allows users to talk to one another and to gain access to the programming language itself. A further improvement was made by Pavel Curtis of Xerox PARC in the early 1990s; he developed an object-oriented programming language for MUDs, which were then called MOOs. The MOO environment is easier to program, with a format closer to natural language, and it allows users to create objects in categories and subcategories. For instance, if I want to make a duck in a MOO, I could start with some generic duck code that causes the duck to quack and swim and could then add my own refinements for my particular duck, like a taste for blue worms or a hostility to geese.
12. I am using the term *constructivism* to indicate an aesthetic enjoyment in making things within a fictional world. Piaget used the same term for the

way in which the child learns by manipulating objects in the world, and Seymour Papert extended Piaget's concept to "constructionism," which is a method of learning by building things; see Papert, *Mindstorms*. An aesthetic experience does not have the same instrumental goals as a learning experience, but it can share the same activities of creating a gestalt (see Iser, *The Act of Reading*). I am grateful to Amy Bruckman for the examples of the spaghetti plate and the fan, presented in her unpublished paper "Identity Workshop" and in her thesis proposal, respectively.

13. There are signs already of an increase in the presence of the constructivist ethos, even in the more conservative world of CD-ROMs. For instance, Steven Spielberg has produced a disc that allows you to create your own movie on an editing table with original raw footage he provides. Many fighting games now begin by asking interactors to build their own avatars. Some even allow you to put together your own combat maze game with custom-made opponents. The World Wide Web is a strongly constructivist environment, offering many sites that allow you to build your own postcard, generate your own poem, construct a gag face, or just appropriate sounds and images for your own use.

14. Laurel, Strickland, and Tow. "Placeholder: Landscape and Narrative in Virtual Environments," 118.

Chapter 6. Transformation

1. *The Norman Conquests* is available as a set of three videotapes of the British television production.

2. The restaurant story is "Welcome to the Tau Cafe" by Alejandro Paris Heyworth. The bus story is "Crosstown" by Michael Murtaugh.

3. Freedom Baird, Mass Transit, available at http://ic.www.media.mit.edu/people/baird/mass_transit/index.html; May 1996.

4. The story of the Brontë's childhood story worlds was first told by Fannie Ratchford.

5. From Brontë, "Tales of the Islanders," *Cosmopolitan Magazine*, October 1911, 611–22.

6. In fact, Zenobia is something like the imperious princess Augusta Geraldine Almeda, the central focus of Emily's imaginary kingdom of Gondal.

7. Brontë, *The Spell: An Extravaganza*, 37.

8. Ratchford, *The Brontës' Web of Childhood*, 242–43.

9. *Caroline Vernon* is dated March 26, 1839. By the end of the year, Charlotte wrote a formal farewell to Angria in the form of a handwritten note addressed to her (imaginary) readers. In the note she announces that she has written "a great many books" showing the same places and people in every different weather and mood, "but we must change, for the eye is tired of the picture so often recurring" (Ratchford, *The Brontës' Web of Childhood*, 149).

10. Winnicott, *Playing and Reality*, 5.

11. "Virtual Reality Conquers Fear of Heights," *New York Times*, 21 June 1995,

C11; "Virtual Reality Used as Phobia Therapy," *Boston Globe,* 30 January 1995, pp. 25–26. The reported research was conducted by Dr. Ralph Lamson at Kaiser Permanente Hospital in San Rafael, California, and by Dr. Barbara Rothbaum of Emory University. See Rathbaum (1995).

12. See Sherry Turkle, *Life on the Screen,* 262–63.
13. "Hollow Pursuits," *Star Trek: The Next Generation* (season 3, episode 169, initial broadcast 30 April 1990), written by Sally Caves, directed by Cliff Bole.
14. Landow, *Hypertext,* 113.
15. For discussions of closure in fiction and poetry see Kermode, *The Sense of an Ending,* and Barbara Hernnstein Smith, *Poetic Closure.* For a related discussion of a narrative as a web of potential stories, full of wrong choices and short circuits, see Brooks, *Reading for the Plot,* especially pp. 90–118. For a discussion of closure in hypertext, see Landau, *Hypertext,* 109–112, Moulthrop, "Containing Multitudes," and Harpold, "Conclusions."
16. Umberto Eco interview, on "The Connection," WGBH Radio, November 6, 1995. According to Eco, a hypertext can never be satisfying because "the charm of a text is that it forces you to face destiny."

Chapter 7. The Cyberbard and the Multiform Plot

1. See Campbell, *The Hero with a Thousand Faces.*
2. Tobias, *Twenty Master Plots.*
3. For the most influential, lucid, and persuasive description of the differences between oral and literate intellectual processes and linguistic habits, see Ong, *Orality and Literacy.*
4. Cited from "A Story as You Like It: Interactive Version," at http://fub46.zedat.fu-berlin.de:8080/~cantsin/queneau_20.html, a World Wide Web version of "Cent mille milliards de poemes" by Raymond Queneau, with text adapted from Motte, *Oulipo,* 156–58. Queneau's experiment was part of the 1960s movement called Ouvroir de Litterature Potentielle (OULIPO). He published the original in the form of a book of ten sonnets with pages cut into one-line strips for interchangeablility of lines. In presenting his system for generating "one hundred thousand billion sonnets," he expressed the belief that "poetry should be made by everyone."
5. Lord, *The Singer of Tales,* 123.
6. Ibid., 100.
7. Propp, *Morphology of the Folktale.*
8. I have simplified Propp's formulas somewhat. See Propp, *Morphology of the Folktale,* 103–5. See Lakoff, "Structural Complexity in Fairy Tales," for an attempt to expand Propp's morphemes into a generative story grammar.
9. Amy Bruckman of MIT's Media Lab has pointed out in an unpublished paper, "The Combinatorics of Storytelling: Mystery Train Interactive" (1990), that even a simple "choose your own adventure" story that provides only two menu choices at each choice-point and keeps its branches to a maximum of five choice-points deep would generate an unmanageable

thirty-two possible endings if it did not loop back on itself and merge branches. If the stories were a more satisfying ten choice-points deep, there would be 1,024 possible endings.

10. Winston, *Artificial Intelligence*, 417.
11. Ibid., 417.
12. Schank, *The Cognitive Computer: On Language, Learning, and Artificial Intelligence*, 84–85. See also Meehan, "Tail-spin."
13. Laurel, *Computers as Theatre*, 135–39.
14. Ryan, *Possible Worlds*, 248–57.
15. For the UNIVERSE system, see Michael Lebowitz,"Creating Characters in a Story-Telling Universe" and "Story-Telling as Planning and Learning."
16. See Lebowitz,"Story-Telling as Planning and Learning," 484, and the discussion in Ryan, *Possible Worlds*, 246; see Ryan generally for a more extensive and technical account of artificial intelligence approaches to (nonparticipatory) storytelling.
17. Kelso, Weyhrauch, and Bates, "Dramatic Presence."
18. For Propp's variants of rescue from pursuit (excerpted and paraphrased here) see *Morphology of the Folktale*, 57–58.
19. I am using the notion of the interactor's saturation here as I presented it in chapter 6, based on Winnicott's description of children's play ending when the situation has absorbed all the emotional charge the child brings to it.
20. See Minsky, *The Society of Mind*, 24.2. Minsky's notion of frames is an engineer's view of what cognitive psychology more commonly calls "schemata." For a more detailed explanation (aimed at the lay reader) of how frames work, see Boden, *Artificial Intelligence and Natural Man*, 305ff. See Gardner, *The Mind's New Science*, 58–59, 126–28, for a summary of the cognitive theory of schemata, including its origin in the work of the philosopher Kant.

Chapter 8. Eliza's Daughters

1. *Chatterbot* is now used as a generic term for talking electronic characters. It was coined by Michael Mauldin, the creator of Julia; see Mauldin, *ChatterBots, TinyMUDs, and the Turing Test*.
2. From Foner, "Entertaining Agents." The punctuation in the dialogues in which Julia appears has been altered here for greater readability.
3. See Turing, "Computing Machinery and Intelligence."
4. In fact, Turing's original design for a test of computer intelligence called for a first round in which judges were asked to distinguish between two interviewees, one of whom was always a woman and the other of whom (unbeknown to the judges) was either a man or a computer emulating a man.
5. In order to promote good character making, I hold a contest every year in my course on interactive fiction writing. The students, who range from MIT freshman to Media Lab graduate students, make their own Eliza-like characters using a template-based authoring system that allows them to

connect key words with multiple responses and to specify something about how the responses are chosen. It is an anti-Turing contest with the prize (in ice cream rather than in stacks of dollar bills) going to the program and the human partner who can *sustain* a coherent conversation for the longest possible time. The contest recognizes that a conversation with Eliza is a collaborative improvisation. The authoring environment, Character-Maker/Conversation, which I designed, was programmed by Jeffrey Morrow (in HyperCard and C) and Matthew Gray (in Perl, with html interface). A Java/html version is under construction, programmed by Wu Yuanqing.

6. Salesman is the creation of Matthew Gray.
7. Colby, *Artificial Paranoia*, 36.
8. Ibid., 75–76.
9. At this point PARRY passed the first part of a classic Turing test, with the psychiatrists focused on detecting paranoia (Turing suggested the first tests should direct people to focus on gender) rather than on detecting a computer.
10. Colby, *Artificial Paranoia*, 55–57.
11. See Colby and Gilberte, *Programming a Computer Model of Neurosis*, for a description of the Neurotic Woman program, and see Boden, *Artificial Intelligence and Natural Man*, 22–63, for a layperson's view of its working. Although Colby was attempting to model the psyche, I see his work, like Weizenbaum's, as essentially a work of electronic fiction.
12. See Schank and Abelson, *Scripts, Plans, Goals and Understanding*.
13. Joseph Bates makes this comparison, citing, for instance, Woody Bledsoe's presidential address to the American Association of Artificial Intelligence in 1985 in which Bledsoe described the dream of building a computer-based agent that could "understand, act autonomously, think, learn, enjoy, hate" and who even "liked to walk and play Ping-Pong, especially with me."
14. I have edited this transcript to make it more readable as a dramatic scene, but I have not distorted the computer's responsiveness. See Bates, Loyall, and Reilly, "An Architecture for Action, Emotion, and Social Behavior."
15. Ortony, A., G. Clore, and A. Collins. *The Cognitive Structure of Emotions*.
16. Bates, Loyall, Reilly, "An Architecture for Action, Emotion, and Social Behavior."
17. The Oz group has increasingly emphasized the creation of personality as a function not just of "features" but of their whole character-building architecture. They distinguish themselves from, for instance, the ALIVE project of MIT by their focus on believable (i.e., artistically inflected) rather than lifelike (i.e., scientifically modeled) characters. See, for instance, Reilly, *Proceedings of the First International Conference on Autonomous Agents*, 114–21.
18. MIT students conduct this conversation on Unix machines within a programmable text editing program called EMACs, a combination word processor and hacker's paradise invented by Richard Stallman.

19. Zippy is the creation of Bill Griffith and distributed by King Features Syndicate.

20. From the scenario for "The Stone Guest," in Oreglia, *The Commedia dell'Arte*, 44.

21. From Andrea Perrucci, Dell'arte rappresentativa, premeditata e all'improvviso, 1699, quoted in Oreglia, *The Commedia dell'Arte*, 119–22.

22. From Placido Adriani, *Selva, or the Miscellany of Comic Conceits*, 1739, cited in Oreglia, *The Commedia dell'Arte*, 14–16 .

23. Ken Perlin and Athomas Goldberg, "Improvisational Actors," presented at Lifelike Computer Characters Conference, Snowbird, Utah, September 1995. The project is being developed at the Media Research Lab of New York University.

24. Bruce Blumberg worked on Silas as part of the ALIVE project, as a graduate student under Pattie Maes. In September 1996 he became an assistant professor at the Media Lab and founded the Synthetic Characters Group. See Blumberg's "Action-Selection in Hamsterdam" and "Old Tricks, New Dogs."

25. See Maes, "How to Do the Right Thing," 291–524, and Blumberg, "Action-Selection in Hamsterdam."

26. Rhodes, "PHISH-Nets." Rhodes suggests that a writer could use the goal architecture to specify personality traits so that, for example, "a lazy character might have a goal to minimize the amount of work he has to do," a goal that would cause him to select different kinds of actions than an energetic or heroic character would select.

27. Forster, *Aspects of the Novel*, 52.

28. Ibid., 54.

29. I am not suggesting remaking the movie *Casablanca* as a multiform story. I am merely using the story patterns we are familiar with in *Casablanca* as a way of thinking about characters in a procedurally authored story world.

30. Toni Dove, "Artificial Changelings: A Work of Responsive Cinema in Progress," MEDIA LAB COLLOQUIUM, October 18, 1995.

31. Dogz and Catz are part of the Computer Petz series of animated figures designed by PF Magic. The Web site for these products features an area in which owners can swap snapshots of their digital pets using a screen capture feature within the program.

Chapter 9. Digital TV and the Emerging Formats of Cyberdrama

1. For the best statement of Negroponte's vision of a world in which much that is now done by "atoms" (or separate physical objects) is transferred to "bits" or electronic representations, see *Being Digital*. Negroponte's work with what we now think of as interactive multimedia dates to the late 1960s in the Architecture Machine Group, which became the foundation of the current MIT Media Lab, founded in 1985.

2. The programming of these new "interactive television" networks will prob-

ably evolve from the formats of current television shows. At first, net-worked gaming—from arcade-style combat, to virtual car racing, to team "Jeopardy"—may receive the most development attention, as viewers and advertisers enjoy the novelty of real-time national and international com-petitions among millions of people. But game shows are basically a form of spectacle, and as we have noticed before, participatory spectacle tends to evolve into narrative. So after the gaming novelty settles down into a few stable genres, the home digital domain may resemble print, film, and tele-vision in focusing on fictional storytelling.

3. We have much to learn about the shift to print during the European Re-naissance, but it seems clear that one effect of the arrival of the printing press was the formation of private libraries and the extension of the sources of information away from the sermon and university lecture and toward in-dividually chosen reading. In Victorian times genteel women were often expected to spend their evenings sewing and being read to. Florence Nightingale found the practice of being read to torture, akin to "lying on one's back, with one's hands tied and having liquid poured down one's throat" (see Murray, *Strong-Minded Women*, 91–92).

4. For the work of the Interactive Cinema Group, see the Davenport articles in the bibliography or http://ic.www.media.mit.edu/.

5. Thorburn, "Television as an Aesthetic Medium."

6. The model of the Greek chorus has been attempted in a project on civil rights history done at Mitsubishi Electric Research Lab in Cambridge, Mass. For a description of Tired of Giving In (1995), designed by Carol Strohecker, Larry Friedlander, and Kevin Brooks, see http://atlantic.merl.com//projects/stories/index.html or http://www.merl.com.

7. For the story of Habitat, see Morningstar and Farmer, "The Lessons of Lu-casfilm's Habitat."

Chapter 10. Hamlet on the Holodeck?

1. For instance, Neil Postman (*Amusing Ourselves to Death*, 331) considers good television more dangerous than bad: "We would all be better off if television got worse, not better. *The A-Team* and *Cheers* are no threat to our public health. *60 Minutes, Eye-Witness News* and *Sesame Street* are."

2. Toni Morrison in speech to the American Writers Congress, October 9, 1981, New York City. See Morrison, "Writers Together," 397.

3. The Mario Brothers games, produced by Nintendo, were developed by Shigeru Miyamoto. As Edward Rothstein of the *New York Times* has pointed out, his games "were something apart: they created a new genre out of deceptively simple ideas. A character moves from left to right across a scrolling universe, jumping, hitting and kicking. The opponents are odd, whimsical creatures in a tricky, puzzle-filled world. The real challenger was Mr. Miyamoto himself; virtuosos match themselves against his inventive-ness."

The *King's Quest* series, with seven games published since 1986 by Sierra On-Line, is designed by Roberta Williams. *King's Quest* follows the adventures of a set of continuing characters—including King Graham, Queen Valanice, and Princess Rosella—in the whimsical Kingdom of Daventry.

4. A commedia troupe could revolve around a particular star actor or actor/director, but the success of the play rested on skillful team work. See Nicole, *The World of Harlequin,* especially pp. 24–39.

5. For a discussion of the novelty of Shakespeare's use of the soliloquy in Hamlet, see Kermode, "Introduction to *Hamlet, Prince of Denmark*," 1139. See Shakespeare's *Richard III* and Marlowe's *Faust* for other examples of the soliloquy form moving from plot exposition to psychological revelation. In all three plays, the soliloquy format enacts the dangerous isolation of the protagonist from the society in which he moves.

6. Psychoanalysis can be viewed as a form of narrative art in which patients retell the story of their life in order to discover a version that allows for a more open-ended future. See Schafer, "Narration in the Psychoanalytic Dialogue."

7. D. H. Lawrence, "Morality and the Novel (1925)," 108–13.

Bibliography

Print

Abelson, Harold, and Gerald Jay Sussman. *Structure and Interpretation of Computer Programs.* Cambridge, MA: MIT Press, 1996.

Abt, Clark C. *Serious Games.* New York: Viking, 1970.

Allen, Woody. "The Kugelmass Episode." In *Side Effects,* 59–78. New York: Ballantine Books, 1981.

Avedon, Elliot M., and Brian Sutton-Smith. *The Study of Games.* Huntington, NY: R. E. Krieger, 1979.

Ayckbourn, Alan. *The Norman Conquests : A Trilogy of Plays.* New York: Grove Press, 1988. Also available on HBO video, set of 3 videotapes, 1980.

Barthes, Roland. *S/Z.* Translated by Richard Miller. New York: Hill and Wang, 1974.

Bates, Joseph. "Virtual Reality, Art, and Entertainment." *Presence: The Journal of Teleoperators and Virtual Environments* 1, no. 1 (1992): 133–38.

Bates, Joseph. "The Role of Emotion in Believable Agents." *Communications of the ACM [Association for Computing Machinery]* 37, no. 7 (July 1994): 122–25.

Bates, Joseph, A. Bryan Loyall, and W. Scott Reilly. "An Architecture for Action, Emotion, and Social Behavior." In *Artificial Social Systems: Fourth European Workshop on Modeling Autonomous Agents in a Multi-Agent World.* Berlin: Springer-Verlag, 1994. Also available via ftp at http://www.cs.cmu.edu/afs/cs.cmu.edu/project/oz/web/papers.html

Baudrillard, Jean. *Simulations.* New York: Semiotext(e) Inc., 1983.

Beardsley, Monroe C. *Aesthetics from Classical Greece to the Present.* New York: Macmillan, 1966.

Benedikt, Michael, ed. *Cyberspace: First Steps.* Cambridge, MA: MIT Press, 1992.

Benjamin, Walter. "The Work of Art in the Age of Mechanical Reproduction (1935)." In *Illuminations,* 665–81. New York: Harcourt Brace Jovanovich, 1968.

Birkerts, Sven. *The Gutenberg Elegies: The Fate of Reading in an Electronic Age.* Boston: Faber & Faber, 1994.

Blumberg, Bruce. "Action-Selection in Hamsterdam: Lessons from Ethology." In *From Animals to Animats: Proceedings of the Third International Conference on the Simulation of Adaptive Behavior,* edited by P. Husbands, D. Cliff, J. A. Meyer, and S. W. Wilson. Cambridge MA: MIT Press, 1994.

Blumberg, Bruce. "Old Tricks, New Dogs: Ethology and Interactive Creatures." Ph.D. dissertation, Media Lab, Massachusetts Institute of Technology, 1996.

Boden, Margaret. *Artificial Intelligence and Natural Man.* New York: Basic Books, 1977.

Bolter, Jay David. *Writing Space: The Computer in the History of Literacy.* Hillsdale, NJ: Lawrence Erlbaum, 1989.

Bordwell, David. *Narration in the Fiction Film.* Madison: University of Wisconsin Press, 1985.

Borges, Jorge Luis. "The Garden of Forking Paths" (1942). In *Ficciones,* 89–115, translated by Anthony Kerrigan. New York: Grove Press, 1962.

Bradbury, Ray. *Fahrenheit 451.* New York: Ballantine Books, 1953.

Brewer, D. S., ed. *Malory: The Morte Darthur.* Evanston, IL: Northwestern University Press, 1970.

Brontë, Charlotte. "Caroline Vernon." In *Legends of Angria,* edited by Fannie Ratchford. London: Folio Society, 1933.

Brontë, Charlotte. "Tales of the Islanders." *Cosmopolitan Magazine,* October 1911, 611–22.

Brontë, Charlotte. *The Spell: An Extravaganza,* edited and with an introduction by George Edwin MacLean. London: Oxford University Press, 1931.

Brooks, Peter. *Reading for the Plot: Design and Intention in Narrative.* New York: Vintage/Random House, 1984.

Brooks, Rodney. "Elephants Don't Play Chess." In *Designing Autonomous Agents: Theory and Practice from Biology to Engineeering and Back,* edited by Pattie Maes, 3–15. Cambridge, MA: MIT Press, 1992.

Bruckman, Amy. "MOOSE Crossing: Creating a Learning Culture." Unpublished paper, 1994 (available via ftp at http://asb.www.media.mit.edu/people/asb).

Bruckman, Amy. "The Combinatorics of Storytelling: Mystery Train Interactive." Unpublished paper, 1990.

Bruckman, Amy. "Identity Workshop: Emergent Social and Psychological Phenomena in Text-Based Virtual Reality." Unpublished paper, 1992. Available via ftp from http://asb.www.media.mit.edu/people/asb.

Bush, Vannevar. "As We May Think." *Atlantic Monthly,* July 1945, 101–8.

Calvino, Italo. *If on a Winter's Night a Traveler.* Translated by William Weaver. New York: Harcourt Brace Jovanovich, 1979.

Campbell, Joseph. *The Hero with a Thousand Faces.* New York: Pantheon Books, 1949, 1965.

Canemaker, John. *Winsor McCay: His Life and Art.* New York: Abbeville Press, 1987.

Card, Orson Scott. *Ender's Game.* New York: Tom Doherty Associates, 1986.

Chatman, Seymour. *Story and Discourse.* Ithaca, NY: Cornell University Press, 1978.

Colby, Kenneth Mark. *Artificial Paranoia: A Computer Simulation of Paranoid Process.* New York: Pergamon Press, 1975.

Colby, Kenneth Mark, and J. P. Gilberte. "Programming a Computer Model of Neurosis." *Journal of Mathematical Psychology* 1 (1964): 405–17.

Cook, David. *A History of Narrative Film.* New York: Norton, 1990.

Coover, Robert. "The End of Books." *New York Times Book Review,* June 21, 1992.

Coover, Robert. "Hyperfiction: Novels for the Computer." *New York Times Book Review,* August 29, 1993.

Crane, Gregory. *Perseus 2.0: Interactive Sources and Studies on Ancient Greece.* New Haven, CT: Yale University Press, 1996.

Daley, A. Stuart. "A Chronology of Wuthering Heights." In *Wuthering Heights: A Norton Critical Edition,* edited by William M. Sale and Richard J. Dunn. New York: Norton, 1990.

Davenport, Glorianna (with A. Pentland, R. Picard, and K. Haase). "Video and Image Semantics: Advanced Tools for Telecommunications." *IEEE [Institute of Electrical and Electronics Engineers] Multimedia* 1, no. 2 (Summer 1994): 73–75.

Davenport, Glorianna. "Still Seeking: Signposts of Things to Come?" *IEEE [Institute of Electrical and Electronics Engineers] Multimedia* (Fall 1995): 9–13.

Davenport, Glorianna. "1001 Electronic Story Nights: Interactivity and the Language of Storytelling." Presentation to the Australian Film Commission's 1996 Language of Interactivity conference, Summer 1996.

Davenport, Glorianna. "Smarter Tools for Storytelling: Are They Just Around the Corner?" *IEEE [Institute of Electrical and Electronics Engineers] Multimedia* 4, no. 1 (1996): 10–14.

Davenport, Glorianna. "Indexes Are Out, Models Are In." *IEEE [Institute of Electrical and Electronics Engineers] Multimedia* 4, no. 3 (Fall 1996).

Davenport, Glorianna. "Seeking Dynamic, Adaptive Story Environments." *IEEE [Institute of Electrical and Electronics Engineers] Multimedia* 1, no. 3 (1994): 9–13.

Davenport, Glorianna, and Larry Friedlander. "Interactive Transformational Environments: The Wheel of Life." In *Contextual Media,* edited by Edward Barrett. Cambridge, MA: MIT Press, 1995.

Deleuze, Gilles, and Felix Guattari. *A Thousand Plateaux: Capitalism and Schizophrenia.* Minneapolis: University of Minnesota Press, 1987.

Donaldson, Peter. "The Shakespeare Interactive Archive: New Directions in Electronic Scholarship on Text and Performance." In *Contextual Media,* edited by Edward Barrett. Cambridge, MA: MIT Press, 1995.

Douglas, J. Yellowlees. "How Do I Stop This Thing? Closure and Indeterminacy in Interactive Narratives." In *Hyper/Text/Theory,* edited by George Landow, 159–88. Baltimore: The Johns Hopkins University Press, 1994.

Dove, Toni, and Michael Mackenzie. "Archeology of a Mother Tongue." In *Immersed in Technology: Art and Virtual Environments,* edited by Mary Anne Moser, 275–380. Cambridge, MA: MIT Press, 1996.

Eco, Umberto. *The Role of the Reader.* Bloomington: Indiana University Press, 1979.

Eco, Umberto. *Travels in Hyperreality.* Orlando, FL: Harcourt Brace Jovanovich, 1986.

Eco, Umberto. *The Open Work.* Cambridge, MA: Harvard University Press, 1989.

Eddings, Joshua. *How Virtual Reality Works.* Emoryville, CA: Ziff Davis, 1994.

Eisenstein, Elizabeth L. *The Printing Revolution in Early Modern Europe.* Cambridge, England: Cambridge University Press, 1983.

Eliot, T. S. *Selected Essays.* London: Faber & Faber, 1963.

Eliot, T. S., ed. *Hamlet* (1919). Edited by Frank Kermode. New York: Harcourt Brace Jovanovich, 1975.

Emmeche, Claus. *The Garden in the Machine.* Princeton, NJ: Princeton University Press, 1994.

Felshin, Sue. "The Athena Language Learning NLP System: A Multilingual NLP System for Conversation-based Language Learning." In *Intelligent Language Tutors: Balancing Theory and Technology,* edited by Melissa Holland and Jonathan Kaplan. Hillsdale, NJ: Lawrence Erlbaum, 1994.

Fine, Gary Alan. *Shared Fantasy: Role-Playing Games as Social Worlds.* Chicago: Chicago University Press, 1983.

Foner, Leonard. "Entertaining Agents: A Sociological Case Study." In *Proceedings of the First International Conference on Autonomous Agents,* edited by W. Lewis Johnson. New York: ACM Press. See also "Agents Memo 93.01, MIT Media Lab" available at http://foner.www.media.mit.edu/people/foner/Julia/.

Forster, E. M. *Aspects of the Novel.* London: Arnold, 1927.

Friedman, Ted. "Making Sense of Software: Computer Games and Interactive Textuality." In *Community in Cyberspace,* edited by Steve Jones. Beverly Hills, CA: Sage Publications, 1994.

Frye, Northrop. *Anatomy of Criticism.* Princeton, NJ: Princeton University Press, 1957.

Gardner, Howard. *The Mind's New Science: A History of the Cognitive Revolution.* New York: HarperCollins, 1986.

Gerin, Winnifred. *Charlotte Brontë: The Evolution of Genius.* Oxford, England: Oxford University Press, 1967.

Gibson, William. *Neuromancer.* New York: Ace Books, 1983.

Gunning, Tom. "An Aesthetic of Astonishment: Early Film and the (In)Credulous Spectator." In *Viewing Positions: Ways of Seeing Film,* edited by Linda Williams, 114–33. New Brunswick, NJ: Rutgers University Press, 1995.

Harpold, Terry. "Conclusions." In *Hyper/Text/Theory,* edited by George Landow, 189–224. Baltimore: The Johns Hopkins University Press, 1994.

Harvey, David. *The Condition of Post Modernity.* Cambridge, MA: Basil Blackwell, 1991.

Hauser, Arnold. *The Social History of Art.* Vol. 2, *Renaissance, Mannerism, Baroque.* New York: Vintage, 1960.

Holland, Norman. *The Dynamics of Literary Response.* Oxford, England: Oxford University Press, 1968.

Holland, Norman N. *Five Readers Reading.* New Haven, CT: Yale University Press, 1975.

Hollander, Anne. *Moving Pictures.* Cambridge, MA: Harvard University Press, 1991.

Huxley, Aldous. *Brave New World.* 1932. Reprint, New York: Bantam, 1953.

Iser, Wolfgang. *The Act of Reading: A Theory of Aesthetic Response,* Baltimore: Johns Hopkins University Press, 1978.

Jay, Martin. "The Apocalyptic Imagination and the Inability to Mourn." In *Rethinking Imagination: Culture and Creativity,* edited by Gillian Robinson and John Rundell, 30–45. London: Routledge, 1994.

Jenkins, Henry. *Textual Poachers: Television Fans and Participatory Culture.* New York and London: Routledge, 1992.

Jenkins, Henry. "The Politics of Fandom." *Harper's* 292, no. 1753 (1996): 15.

Johnson, Crockett. *Harold and the Purple Crayon.* New York: Scholastic Book Services, 1955.

Joyce, Michael. *Of Two Minds: Hypertext Pedagogy and Poetics.* Ann Arbor: University of Michigan Press, 1995.

Kelso, Margaret Thomas, Peter Weyhrauch, and Joseph Bates. "Dramatic Presence." *PRESENCE: The Journal of Teleoperators and Virtual Environments* 2, no. 1 (1993). Also available via ftp at http://www.cs.cmu.edu/afs/cs.cmu.edu/project/oz/web/papers.html.

Kermode, Frank. *The Sense of an Ending.* New York: Oxford University Press, 1967.

Kermode, Frank. "Introduction to *Hamlet, Prince of Denmark.*" In *The Riverside Shakespeare,* edited by G. Blakemore Evans, 1135–1140. Boston: Houghton Mifflin, 1974.

Kern, Stephen. *The Culture of Time and Space 1880–1918.* Cambridge, MA: Harvard University Press, 1983.

Krauss, Lawrence. *The Physics of Star Trek.* New York: Harper Perennial, 1996.

Krizanc, John. *Tamara.* Toronto: Stoddard Publishing, 1989.

Lakoff, George. "Structural Complexity in Fairy Tales." *The Study of Man* 1 (1972): 128–50.

Landow, George. *Hypertext: The Convergence of Contemporary Critical Theory and Technology.* Baltimore: The Johns Hopkins University Press, 1992.

Langer, Suzanne. *Feeling and Form.* New York: Scribner, 1953.

Laurel, Brenda. *Computers as Theatre.* Reading, MA: Addison-Wesley, 1993.

Laurel, Brenda, Rachel Strickland, and Rob Tow. "Placeholder: Landscape and Narrative in Virtual Environments." *Computer Graphics: A Publication of ACM SIGGRAPH* 28, no. 2 (1994): 118–26.

Lawrence, D. H. "Morality and the Novel (1925)." In *D. H. Lawrence: Selected*

Literary Criticism, edited by Anthony Beal, 108–113. New York: Viking Press, 1966.

Lawrence, D. H. "Why the Novel Matters." In *D. H. Lawrence: Selected Literary Criticism,* edited by Anthony Beal. New York: Viking Press, 1966.

Lebling, P. David, Marc S. Blank, and Timothy A. Anderson. "Zork: A Computerized Fantasy Simulation Game." *IEEE Computer* 12, no. 4 (April 1979): 51–59.

Lebowitz, Michael. "Creating Characters in a Story-Telling Universe." *Poetics* 13 (1984): 171–94.

Lebowitz, Michael. "Story-Telling as Planning and Learning." *Poetics* 14 (1985): 483–502.

LeGuin, Ursula. *Lathe of Heaven.* New York: Scribner, 1971.

Lightman, Alan. *Einstein's Dreams.* New York: Pantheon, 1993.

Lippman, Andrew. "Movie Maps." *ACM SIGGRAPH* 14 (1980): 3.

Lippy, Tod. "Harold Ramis on Groundhog Day." *Scenario: The Magazine of Screenwriting Art* 1, no. 2 (1995): 53.

Lippy, Tod, interviewer. "A Talk with Danny Rubin." *Scenario: The Magazine of Screenwriting Art* 1, no. 2 (1995): 49–52, 183–87.

Lodge, David. *The Art of Fiction: Illustrated from Classic and Modern Texts.* New York: Viking, 1992.

Lord, Albert B. *The Singer of Tales.* Cambridge, MA: Harvard University Press, 1960.

Maes, Pattie. "How to Do the Right Thing," *Connection Society Journal* 1, no. 3 (1989): 291–524.

Maes, Pattie. "Artificial Life Meets Entertainment: Lifelike Autonomous Agents." *Communications of the ACM [Association for Computing Machinery]: Special Issue on New Horizons of Commercial and Industrial: Artificial Intelligence* 38, no. 11 (1995): 108–114. Also available via ftp at http://pattie.www.media.mit.edu/people/pattie/cv.html#publications.

Maes, Pattie, ed. *Designing Autonomous Agents: Theory and Practice from Biology to Engineering and Back.* Cambridge, MA: MIT Press, 1992.

Maes, P., T. Darrell, B. Blumberg, and A. Pentland. "The ALIVE System: Wireless, Full-Body Interaction with Autonomous Agents." *ACM [Association for Computing Machinery] Multimedia Systems: Special Issue on Multimedia and Multisensory Virtual Worlds,* Spring 1996. Also available via ftp at http://pattie.www.media.mit.edu/people/pattie/cv.html#publications.

Malone, Stuart A., and Sue Felshin. "GLR Parsing for Erroneous Input." In *Generalized LR Parsing,* edited by Masaru Tomita. Boston: Kluwer, 1991.

Martin, Wallace. *Recent Theories of Narrative.* Ithaca, NY: Cornell University Press, 1986.

Mast, Gerald, and Bruce F. Kawin. *A Short History of the Movies.* Boston: Allyn and Bacon, 1996.

Mauldin, Michael. "ChatterBots, TinyMUDs, and the Turing Test: Entering the Loebner Prize Competition." Paper presented at the Twelfth National

Conference on Artificial Intelligence, Menlo Park, CA, 1994. Available on-line at : http://fuzine.mt.cs.cmu.edu/mlm/aaai94.html.

McCloud, Scott. *Understanding Comics: The Invisible Art.* New York: Harper-Collins, 1994.

McCorduck, Pamela. *Machines Who Think: A Personal Inquiry into the History and Prospects of Artificial Intelligence.* New York: Freeman, 1979.

McLuhan, Marshall. *The Gutenberg Galaxy: The Making of Typographic Man.* New York: New American Library, 1969.

McLuhan, Marshall. *Understanding Media.* New York: McGraw-Hill, 1964.

McLuhan, Marshall, and Quentin Fiore. *The Medium is the Massage: An Inventory of Effects.* New York: Random House, 1967.

Meehan, James. "Tail-spin." In *Inside Computer Understanding,* edited by Roger Schank, 197–225. Hillsdale, NJ: Lawrence Erlbaum, 1981.

Minsky, Marvin. *The Society of Mind.* New York: Simon & Schuster, 1986, 1988.

Morgenstern, Douglas, and Janet Murray. "Tracking the Missing Biologist." *Humanities* 16, no. 5 (1995): 33–38.

Morningstar, Chip, and F. Randall Farmer. "The Lessons of Lucasfilm's Habitat." In *Cyberspace: First Steps,* edited by Michael Benedikt. Cambridge, MA: MIT Press, 1992.

Morrison, Toni. "Writers Together." *The Nation,* October 24, 1981: 396–97, 412.

Morse, Margaret. "Nature Morte: Landscape and Narrative in Virtual Environments." In *Immersed in Technology: Art and Virtual Environments,* edited by Mary Anne Moser, 195–232. Cambridge, MA: MIT Press, 1996.

Moser, Mary Anne, ed. *Immersed in Technology: Art and Virtual Environments.* Cambridge, MA: MIT Press, 1996.

Motte, Warren, F., Jr. *Oulipo: A Primer of Potential Literature.* Lincoln: University of Nebraska Press, 1986.

Moulthrop, Stuart. "Containing Multitudes: The Problem of Closure in Interactive Fiction." *Association for Computers and the Humanities Newsletter* 10 (1988): 1, 7.

Moulthrop, Stuart. "Toward a Paradigm for Reading Hypertexts: Making Nothing Happen in Hypermedia Fiction." In *Hypertext/Hypermedia Handbook,* edited by E. Berk and J. Devlin, 65–78. New York: McGraw Hill, 1991.

Moulthrop, Stuart. "Rhizome and Resistance: Hypertext and the Dreams of a New Culture." In *Hyper/Text/Theory,* edited by George Landow, 299–319. Baltimore: The Johns Hopkins University Press, 1994.

Murray, Janet H. *Strong-Minded Women and Other Lost Voices from Nineteenth-Century England.* New York: Pantheon, 1982.

Murray, Janet H. "Courtship and the English Novel: Feminist Readings in the Fiction of George Meredith." In *Harvard Dissertations in American and English Literature Series,* edited by Stephen Orgel. New York: Garland, 1987.

Murray, Janet H. "Emerging Genres of Interactive Videodiscs for Language Instruction." In *Multimedia and Language Learning,* edited by James Noblitt. Chapel Hill, NC: Institute for Academic Computing, 1990.

Murray, Janet H. "Anatomy of a New Medium: Literary and Pedagogic Uses of Advanced Linguistic Computer Structures." *Computers and the Humanities* 25, no. 1 (1991): 1–14.

Murray, Janet H., ed. *Miss Miles (1890), by Mary Taylor.* New York: Oxford University Press, 1991.

Murray, Janet H. "Restructuring Space, Time, Story, and Text in Advanced Multimedia Learning Environments." In *Multimedia, Hypermedia, and the Social Construction of Knowledge,* edited by Edward Barrett, 319–45. Cambridge, MA: MIT Press, 1992.

Murray, Janet H. "Lessons Learned from the Athena Language Learning Project: Using Natural Language Processing, Graphics, Speech Processing, and Interactive Video for Communication-Based Language Learning." In *Intelligent Language Tutors: Balancing Theory and Technology,* edited by Melissa Holland and Jonathan Kaplan. Hillsdale, NJ: Lawrence Erlbaum, 1994.

Murray, Janet H. "The Pedagogy of Cyberspace: Teaching a Course in Reading and Writing Interactive Fiction." In *Contextual Media,* edited by Edward Barrett. Cambridge, MA: MIT Press, 1994.

Murray, Janet, and Myra Stark, eds. *The Englishwoman's Review of Social and Industrial Questions 1866–1910.* 41 vols. New York: Garland Publishing, 1980–1984.

Nelson, Theodor Holm. *Literary Machines. 93.1* Sausalito CA: Mindful Press, 1992.

Newcomb, Horace, and Paul M. Hirsch. "Television as a Cultural Forum." In *Television: The Critical View,* edited by Horace Newcomb. Oxford, England: Oxford University Press, 1994.

Nicoll, Allardyce. *The World of Harlequin.* Cambridge, UK: Cambridge University Press, 1963.

Niesz, Anthony J., and Norman Holland. "Interactive Fiction." *Critical Inquiry* 11 (1984): 110–29.

Ong, Walter J. *Orality and Literacy: The Technologizing of the Word.* New York: Methuen, 1982.

Oreglia, Giacomo. *The Commedia dell'Arte.* London: Methuen, 1968.

Ortony, A., G. Clore, and A. Collins. *The Cognitive Structure of Emotions.* Cambridge, UK: Cambridge University Press, 1988.

Papert, Seymour. *Mindstorms: Children, Computers, and Powerful Ideas.* New York: Basic Books, 1980.

Pavic, Milorad. *Dictionary of the Khasars: A Lexicon Novel in 100,000 Words.* Translated by Christina Pribicevic-Zoric. New York: Knopf, 1988.

Pinsky, Robert. "A Brief Description of Mindwheel." *New England Review and Bread Loaf Quarterly* 10, no. 1 (1987): 64–67.

Pinsky, Robert, and P. Michael Campbell. "Mindwheel: A Game Sesson." *New England Review and Bread Loaf Quarterly* 10, no. 1 (1987): 70–75.

Plato. *The Republic of Plato.* Translated by Francis MacDonald Cornford. Oxford, UK: Oxford University Press, 1970.

Postman, Neil. *Amusing Ourselves to Death: Public Discourse in the Age of Show Business.* New York: Viking, 1985.

Propp, Vladimir. *Morphology of the Folktale.* 2nd ed. Translated by Laurence Scott. Austin: University of Texas Press, 1928, 1968.

Queneau, Raymond. *Cent Mille Milliards de Poemes.* Paris: Gallimard, 1961.

Ratchford, Fannie Elizabeth. *The Brontës' Web of Childhood.* Columbia University Press, 1941.

Reeves, Byron, and Clifford Nass. *The Media Equation: How People Treat Computers, Television, and New Media Like Real People and Places.* New York: Cambridge University Press, 1996.

Reilly, W. Scott. "A Methodology for Building Believable Social Agents." *Proceedings of the First International Conference on Autonomous Agents, Marina del Rey, CA,* edited by W. Lewis Johnson, 114–121. New York: ACM Press, 1997.

Rheingold, Howard. *Virtual Community: Homesteading on the Electronic Frontier.* Reading, MA: Addison-Wesley, 1993.

Rhodes, Bradley James. "PHISH-Nets: Planning Heuristically in Situated Hybrid Networks," MA thesis, MIT Media Arts and Sciences, Massachusetts Institute of Technology, September 1996.

Robinson, Gillian, and John Rundell. *Rethinking Imagination: Culture and Creativity.* London: Routledge, 1994.

Rothbaum, Barbara, Larry F. Hodges, Rob Kooper, Dan Opdyke, and James S. Williford. "Effectiveness of Computer-Generated (Virtual Reality) Graded Exposure in the Treatment of Acrophobia." *American Journal of Psychiatry* 52, no. 4 (April 1995): 626–40.

Rothstein, Edward. Technology column. *New York Times,* November 25, 1996, p. D5.

Rubin, Danny. "Groundhog Day (draft screenplay)." *Scenario: The Magazine of Screenwriting Art,* Spring 1995, 6–47.

Rubin, David C. *Memory in Oral Traditions: The Cognitive Psychology of Epic, Ballads, and Counting-out Rhymes.* New York: Oxford University Press, 1995.

Ryan, Marie-Laure. *Possible Worlds: Artificial Intelligence and Narrative Theory.* Bloomington: Indiana University Press, 1991.

Schafer, Roy. "Narration in the Psychoanalytic Dialogue." In *On Narrative,* edited by W. J. T. Mitchell. Chicago: University of Chicago Press, 1980.

Schank, Roger C. (with Peter G. Childers). *The Cognitive Computer.* Reading, MA: Addison-Wesley, 1984.

Schank, Roger C., and R. P. Abelson. *Scripts, Plans, Goals and Understanding.* Hillsdale, NJ: Lawrence Erlbaum, 1977.

Schwartz, Delmore. "In Dreams Begin Responsibilities." In *In Dreams Begin Responsibilities and Other Stories.* New York: New Directions, 1937, 1978.

Smith, Barbara Hernnstein. *Poetic Closure: A Study of How Poems End.* Chicago: University of Chicago Press, 1968.

Smith, Winifred. *The Commedia dell'Arte.* New York: Benjamin Blom, 1964.

Stone, Allucquere Rosanne. *The War of Desire and Technology at the Close of the Mechanical Age.* Cambridge, MA: MIT Press, 1995.

Thorburn, David. "Television as an Aesthetic Medium." *Critical Studies in Mass Communication* 4 (1987): 161–73.

Thorburn, David. "Interpretation and Judgment: A Reading of *Lonesome Dove.*" *Critical Studies in Mass Communication* 10 (1993): 113–127.

Thorburn, David. "Television Melodrama." In *Television: The Critical View,* edited by Horace Newcomb, 537–50. Oxford, England: Oxford University Press, 1994.

Tobias, Ronald B. *Twenty Master Plots (And How to Build Them).* Cincinnati, OH: Writer's Digest Books, 1993.

Turing, Alan. "Computing Machinery and Intelligence." In *Computers and Thought,* edited by E. A. Feigenbaum and J. Feldman, 11–35. New York: McGraw-Hill, 1950; reprint 1963.

Turkle, Sherry. *The Second Self: Computers and the Human Spirit.* New York: Simon & Schuster, 1984.

Turkle, Sherry. *Life on the Screen: Identity in the Age of the Internet.* New York: Simon & Schuster, 1995.

Turner, Victor. *The Ritual Process: Structure and Antistructure.* Chicago: Aldine, 1966.

Wagner, Jane. *The Search for Signs of Intelligent Life in the Universe.* New York: Harper & Row, 1985.

Watt, Ian. *The Rise of the Novel.* Berkeley: University of California Press, 1957.

Weizenbaum, Joseph. "ELIZA: A Computer Program for the Study of Natural Language Communication Between Man and Machine." *Communications of the ACM [Association for Computing Machinery]* 9 (1966): 36–45.

Weizenbaum, Joseph. *Computer Power and Human Reason.* New York: Freeman, 1976.

Weyhrauch, Peter. "Guiding Interactive Drama." Dissertation, School of Computer Science, Carnegie Mellon University, 1997.

Wiener, Norbert. *Cybernetics: Control and Communication in the Animal and the Machine.* 2nd ed. Cambridge, MA: MIT Press, 1962.

Winnicott, D. W. *Playing and Reality.* New York: Routledge, 1971.

Winston, Patrick Henry. *Artificial Intelligence.* 2nd ed. Reading, MA: Addison-Wesley, 1984.

Wolf, Gary. "The Curse of Xanadu." *Wired,* June 1995, 137–52, 194–202.

Film and TV

Across the Sea of Time. Directed by Stephen Low, written by Andrew Gellis. Sony New Technologies. IMAX format 3-D film, 1995.

Arrivée d'un Train à la Ciotat, L'. Directed by Auguste Lumière and Louis Lumière. Silent film, 1895.

Babylon 5. Created by J. Michael Straczynski. Time Warner Corporation. Television series, debut January 1994.

Back to the Future. Written by Robert Zemeckis and Bob Gale, directed by Robert Zemeckis. Universal Pictures. Film, 1985.

Back to the Future Part II. Written by Robert Zemeckis and Bob Gale, directed by Robert Zemeckis. Universal Pictures. Film, 1989.

"Better Than Life." Episode 8 of *Red Dwarf.* Written by Rob Grant and Doug Naylor, produced and directed by Ed Bye. Paul Jackson Productions. British television series, shown in the United States on public television, initial broadcast September 13, 1988.

Casablanca. Directed by Michael Curtiz, screenplay by Julius J. Epstein, Philip G. Epstein, and Howard Koch (from the play "Everybody Goes to Rick's," by Murray Burnett and Joan Alison). Warner Brothers. Film, 1942.

"Cathexis." Episode 12 of *Star Trek: Voyager,* (Season 1, initial broadcast on May 1, 1995). Written by Brannon Braga, directed by Kim Friedman. Television program.

Dr. Quinn, Medicine Woman. Created by Beth Sullivan. The Sullivan Company. Television series, debut 1992.

Duck Amuck. Created by Chuck Jones. Warner Brothers. Animated film, 1951.

ER. Created by Michael Crichton. Warner Brothers/Constant C Productions/Amblin Entertainment. Television series, debut 1994.

Gertie the Dinosaur. Created by Winsor Z. McCay. Animated Film, 1914.

Groundhog Day. Written by Danny Rubin and Harold Ramis, edited and directed by Harold Ramis. Columbia Pictures. Film, 1993.

"Hollow Pursuits." Episode 169 of *Star Trek: The Next Generation* (Season 3, initial broadcast April 30, 1990). Written by Sally Caves, directed by Cliff Bole. Television program.

Homicide: Life on the Street. Created by Barry Levinson. MCEG Sterling/NBC Productions/Baltimore Pictures. Television series, debut 1993.

It's a Wonderful Life. Directed by Frank Capra, written by Frank Capra, Frances Goodrich, Albert Hackett, Philip Van Doren Stern, and Jo Swerling. RKO Radio Pictures. Film, 1946.

Lawnmower Man. Written and directed by Britt Leonard, written and produced by Gimel Everett. New Line Cinema. Film, 1992.

"Learning Curve." Episode 15 of *Star Trek: Voyager* (Season 1, initial broadcast on May 22, 1995). Written by Ronald Wilkerson and Jean Louise Matthias, directed by David Livingston. Television program.

"Parallels." Episode 163 of *StarTrek: The Next Generation* (Season 7, initial broadcast on November 27, 1993). Written by Brannon Braga, directed by Robert Wiemer. Television program.

"Persistence of Vision." Episode 23 of *Star Trek: Voyager* (Season 2, initial broadcast on October 30, 1995). Written by Jeri Taylor, directed and produced by James L. Conway. Television program.

Rashomon. Directed by Akira Kurosawa, written by Shinobu Hashimoto and

Akira Kurosawa from stories by Ryunosuke Akutagawa. Daiei (Japan). Film, 1950.

Tek War. Created by William Shatner. Cardigan Productions. Television series, debut 1994.

Wings of Courage. Directed by Jean-Jacques Annaud, written by Jean-Jacques Annaud and Alan Godard. Sony Pictures Classics. IMAX format 3-D film, 1995.

Witness. Directed by Peter Weir, written by William Kelley, Earl Wallace, and Pamela Wallace. Paramount. Film, 1985.

Digital Works

A la rencontre de Philippe. Created by Gilberte Furstenberg, Janet H. Murray, Stuart A. Malone, and Ayshe Farman-Farmaian. New Haven, CT: Yale University Press. Videodisc program for Macintosh environment, 1993.

Adventure. Created by William Crowther and Don Woods. Text adventure game available on Unix systems at many universities and research labs, 1976.

Afternoon. Written by Michael Joyce. Cambridge, MA: Eastgate Systems. Storyspace hypertext narrative, 1987.

Civilization (also called *Sid Meier's Civilization*). Created by Sid Meier. Microprose. Simulation game for personal computers, first version 1991.

Computer Petz. Created by Adam Frank and Ben Resner. San Francisco: PF Magic. Series of animated interactive characters for personal computers, introduced 1995.

Crime Story. Created by Tom Arriola. Quest Interactive Media. Web serial (available on the World Wide Web at http://www.quest.net.crime), debut January 15, 1995.

Deadline. Created by Marc Blank. Infocom. Text-based computer adventure game for personal computers, 1982. (Available from Activision.)

Dickens Web, The. Created by George Landow. Cambridge, MA: Eastgate Systems. Storyspace hypertext, 1992.

Doom. Created by John Carmack and John Romeo. Id software. 3D fighting game for arcade and personal computers, original version 1993.

"In Memoriam" Web, The. Created by George Landow and Jon Lanestedt. Cambridge, MA: Eastgate Systems. Storyspace hypertext, 1992.

King's Quest. Created by Roberta Williams. Sierra On-Line. Puzzle game series, begun 1986.

Mad Dog McCree. Albuquerque, NM: American Laser Games. Videogame, 1990.

Mortal Kombat III. Created by Ed Boon and John Tobias. Acclaim. Arcade and videogame series, 1995.

Myst. Created by Rand Miller and Robyn Miller. Broderbund. CD-ROM puzzle game for personal computers, 1993.

Nights into Dreams. Sega Saturn. 3-D adventure game, 1996.

Planetfall. Created by Steve Meretzky. Infocom. Text-based puzzle game for personal computer, 1983. (Available from Activision.)

Quake. Created by John Carmack and Michael Abrash. Id software. Internet and CD-ROM 3-D fighting game, 1996.

7th Guest. Created by Rob Landeros and Graeme Devine. Trilobyte. CD-ROM puzzle game, 1993.

SimCity. Created by Will Wright. Maxis. First in a series of simulation programs for Macintosh and PC, 1987.

Spot, The. Created by Scott Zakarin. AMCY Network. Serial story on the World Wide Web at http://www.thespot.com, debut 1995.

Star Trek The Next Generation: A Final Unity. Spectrum Holobyte. Puzzle game for personal computer, 1995.

Star Trek The Next Generation: Interactive Technical Manual. New York: Simon & Schuster Interactive. CD-ROM "virtual tour" of Starship *Enterprise,* 1994.

Star Wars: Rebel Assault. Created by Vince Lee. LucasArts. CD-ROM fighting game, first in series, 1993.

Star Wars: TIE Fighter. Created by Larry Holland and Ed Killham. LucasArts. CD-ROM fighting and battle simulation game, 1994.

Star Wars: X-Wing. Created by Larry Holland and Ed Kilham. LucasArts. CD-ROM fighting and battle simulation game, 1993.

System-D: Writing Assistant for French. Created by James Noblitt. Boston: Heinle and Heinle. Computer software, 1993.

Taz in Escape from Mars. Sega. Fighting game, 1994.

Tetris. Designed by Alexey Pazhitnov in 1985. Spectrum Holobyte. Graphic puzzle game for videogame players and personal computers, 1987.

Uncle Buddy's Phantom Funhouse. Created by John McDaid. Cambridge, MA: Eastgate Systems. Software and audiotape hypermedia novel, 1983.

Victory Garden. Written by Stuart Moulthrop. Cambridge, MA: Eastgate Systems. Storyspace hypertext narrative, 1992.

XMEN 2: Clone Wars. Videogame for Sega Genesis system, 1995.

Zork. Created by Mark S. Blank, P. David Lebling, and Timothy A. Anderson (mainframe version, 1977). Released by Infocom for personal computers, 1980. Available from Activision. Text-based adventure game.

Index

Deadline game, 62–63
Defense Department, 80
Deleuze, Gilles, 132
"Demon," 78
Diamond Age, The (Stephenson), 121–22
Dickens, Charles, 22, 29, 73, 175
Dictionary of the Khazars (Pavic), 37, 56
Digital environments: as encyclopedic, 83–90;
    essential properties of, 71–90; as participa-
    tory, 74–79; as procedural, 71–74; as spatial,
    79–83. *See also* Cyberdrama; Cyberspace;
    Electronic fiction; Electronic games; Hyper-
    text; Hypertext fiction
Disneyworld, 45
Dr. Quinn: Medicine Woman, 265
"Dogz," 244, 245
Don Quixote (Cervantes), 29, 97–98, 103–104,
    137
Doom game, 62, 145–46
Dove, Toni, 104–105, 244
Drama. *See* Cyberdrama; Theater; and specific
    plays and playwrights
Duck Amuck, 104
Dungeons and Dragons, 42, 79, 82
Dungeons and Dragons tabletop game, 82

Eastgate Systems, 57
Eco, Umberto, 103, 178
Edge of Intention, 232, 234–35
Educational computing, author's experiences
    in, 5–8
Einstein's Dreams (Lightman), 34–35
Electronic fiction: authorship of chatterbots,
    219–22, 281–82; class in writing, 9, 219;
    frame-based authoring system, 208–13; ju-
    venilia stage of, 279–80; kaleidoscopic sto-
    ries, 159–62, 175–80. *See also* Hypertext
    fiction; Participatory narrative; Procedural
    authorship
Electronic games: and active creation of belief,
    111–12; avatars in, 113; compared with sto-
    ries, 140; computer skill games, 144–45;
    contest story in, 145–47; dramatic story-
    telling in, 51–55; exploration of borders of,
    105; fears about, 21–22; fighting games, 54,
    146–47, 198; film techniques in, 53–54; first
    graphics-based games, 80; fixation on, 175;
    games into stories, 140–42; as journey sto-
    ries, 139–40; "losing" endings of, 141–42;
    maze-based games, 51–52, 105, 108,
    130–32; morphemes in, 197–98; naviga-
    tional pleasures of, 129; plot events in, 102;
    plots of losing game, 143; plots of symbolic
    action of, 142; puzzle games and quests,
    52–53, 108–109; and refused closure, 173;
    story line in, 197–98; substitution system
    in, 198; suspense in, 135; as symbolic dra-
    mas, 142–45; visitor role in, 108–109;
    "winning" ending of, 141. *See also* specific
    games

Electronic narrative. *See* Hypertext fiction;
    Narrative; Participatory narrative; Proce-
    dural authorship
Eliot, T. S., 93
ELIZA computer program, 69, 72–74, 76–77,
    106, 222
Eliza (computer-based character), 68–74, 190,
    214, 218, 224, 233–34
Eliza effect, 224
Elizabethan drama, 145, 276–77, 278, 280. *See
    also* Shakespeare, William
Elseworlds comic books, 40
Emergence as animation, 239–47
Emergent behavior, 239–40
Empire Strikes Back, The, 265
Enactment as transformational experience,
    170–73
Encyclopedic expectation, 84
Encyclopedic property: of digital environ-
    ments, 83–90; as handicap, 87, 89–90
Entertainment industry, 252
Entertainment products: emerging cyber-
    drama, 271–72; future projections of,
    252–72; hyperserial, 253–58, 260; mobile
    viewer movies, 258–62; role-playing in au-
    thored world, 266–71; virtual places and fic-
    tional neighborhoods, 262–65
Epcot Center, 45
ER, 157, 162, 255–56, 257, 260, 262
ESG. *See* Experimental Study Group (ESG)
"Evening" (Molenaar), 161, 259
Exhibitionism, 121
Experimental Study Group (ESG), 5

Fahrenheit 451 (Bradbury), 19–21, 99
Fan culture, on Internet, surrounding televi-
    sion series, 41–42, 84–85
"Farewell to Angria" (Brontë), 167
Faulkner, William, 91, 111, 256–57
Fears: about computers, 7–8, 22–24; about
    movies, 18–19, 26; about new representa-
    tional technologies, 17–24, 26, 98; about
    simulations, 103; about television, 19–21;
    about videogames, 21–22
Feeling and Form (Langer), 100–101
Feminism, 4, 121
Fiction. *See* Hypertext fiction; Novels
Figaro (Mozart), 277
Fighting games, 54, 146–47, 198. *See also* spe-
    cific games
Film art digital textbook, 6, 7
Films: animated films, 104, 105–106; cartoons,
    104; computer as dehumanizing in, 23–24;
    early films, 65–66; elements of filmic story-
    telling, 66, 119–20; fears about film
    medium, 18–19, 21, 26; feely theater in
    Huxley's *Brave New World*, 18–19, 21, 26,
    273; IMAX technology for, 44–49, 120–21;
    journey story in, 137; juxtaposition in, 160;
    as linear media, 79; mobile viewer movies of